Urban Social Sustainability

This ground-breaking volume raises radical critiques and proposes innovative solutions for social sustainability in the built environment. *Urban Social Sustainability* provides an in-depth insight into the discourse and argues that every urban intervention has a social sustainability dimension that needs to be taken into consideration, and incorporated into a comprehensive and cohesive 'urban agenda' that is built on three principles of recognition, integration, and monitoring. This should be achieved through a dialogical and reflexive process of decision-making. To achieve sustainable communities, social sustainability should form the basis of a constructive dialogue and be interlinked with other areas of sustainable development. This book underlines the urgency of approaching social sustainability as an urban agenda and goes on to make suggestions about its formulation.

Urban Social Sustainability consists of original contributions from academics and experts within the field and explores the significance of social sustainability from different perspectives. Areas covered include urban policy, transportation and mobility, urban space and architectural form, housing, urban heritage, neighbourhood development, and urban governance. Drawing on case studies from a number of countries and world regions the book presents a multifaceted and interdisciplinary understanding from social sustainability in urban settings, and provides practitioners and policy makers with innovative recommendations to achieve more socially sustainable urban environment.

M. Reza Shirazi is Marie Skłodowska-Curie Global Fellow at the Institute of Urban and Regional Development (IURD), UC Berkeley, and Senior Research Fellow at the School of the Built Environment, Oxford Brookes University, UK. His research and teaching interests lie in the field of urban justice, discourse studies, neighbourhood development, citizen participation, phenomenology of the built environment, sociocultural sustainability, and architectural and urban transformation in the MENA (Middle East and North Africa) countries. Reza's recent books include: *Contemporary Architecture and Urbanism in Iran: Tradition, Modernity, and the Production of 'Space-In-Between'* (Springer, 2018), *Citizens' Participation in Urban Planning and Development in Iran; Challenges and Opportunities* (Routledge, 2017); and *Towards an Articulated Phenomenological Interpretation of Architecture: Phenomenal Phenomenology* (Routledge, 2014).

Ramin Keivani is Professor of International Land Policy and Urban Development and Research Lead at the School of the Built Environment, Oxford Brookes University. He is an urbanist with a particular research interest in low-income and affordable housing policy, urban social sustainability, and the impact of land and property markets on urban development and urban equity in both developed and transition economies. He has worked on projects across the globe including UK, central/eastern Europe, the Middle East, India, Brazil, China and Asia Pacific, and southern Africa. He is the founding and managing editor of the *International Journal of Urban Sustainable Development*.

Routledge Studies in Sustainability

Sustainability Transitions in South Africa
Edited by Najma Mohamed

Strongly Sustainable Societies
Organising Human Activities on a Hot and Full Earth
Edited by Karl Johan Bonnedahl and Pasi Heikkurinen

Managing Complexity, Earth Systems and Strategies for the Future
Walter R. Erdelen and Jacques G. Richardson

Measuring Intangible Values
Rethinking How to Evaluate Socially Beneficial Actions
Marie Harder and Gemma Burford

Education for Sustainable Human and Environmental Systems
From Theory to Practice
Edited by Will Focht, Michael A. Reiter, Paul A. Barresi and Richard C. Smardon

The Politics of Sustainability in the Arctic
Reconfiguring Identity, Space, and Time
Edited by Ulrik Pram Gad and Jeppe Strandsbjerg

Global Planning Innovations for Urban Sustainability
Edited by Sébastien Darchen and Glen Searle

The Question of Limits
A Historical Perspective on the Environmental Crisis
Christian Marouby

Urban Social Sustainability
Theory, Policy and Practice
Edited by M. Reza Shirazi and Ramin Keivani

www.routledge.com/Routledge-Studies-in-Sustainability/book-series/RSSTY

Urban Social Sustainability

Theory, Policy and Practice

Edited by M. Reza Shirazi and Ramin Keivani

Routledge
Taylor & Francis Group
LONDON AND NEW YORK

First published 2019
by Routledge
2 Park Square, Milton Park, Abingdon, Oxon OX14 4RN

and by Routledge
52 Vanderbilt Avenue, New York, NY 10017

First issued in paperback 2020

Routledge is an imprint of the Taylor & Francis Group, an informa business

British Library Cataloguing-in-Publication Data
A catalogue record for this book is available from the British Library

Library of Congress Cataloging-in-Publication Data
A catalog record has been requested for this book

ISBN 13: 978-0-367-66203-5 (pbk)
ISBN 13: 978-1-138-06938-1 (hbk)

Typeset in Bembo
by Wearset Ltd, Boldon, Tyne and Wear

Contents

Figures

Tables

Contributors

Marco Adelfio is a researcher in the Architecture and Civil Engineering Department of Chalmers University, Sweden. His research focuses on multi-stakeholder processes and driving forces contributing to compact, mixed-use and socially sustainable urban/suburban development. He has expertise in mixing qualitative and quantitative methods, assessment of best practice transferability and use of geographical information systems for co-production of knowledge.

Annika Airas is a post-doctoral fellow in the Urban Studies program at Simon Fraser University in Vancouver, Canada. Her recent comparative PhD research focused on the historical distinctiveness of urban waterfront redevelopments in Finland and Canada, outlined in her recent chapter 'Towards homogeneous waterfronts? Historical woodworking waterfronts in transition', published in *Waterways and the Cultural Landscape* (Routledge, 2018). Her continuing research examines redevelopment, planning and policy, particularly in relation to rapidly changing, urban, built environments and social activities.

Mark Davidson is an urban geographer whose research examines the relationship between cities and politics, and the drivers of contemporary urban change. Mark has an international reputation for his research on gentrification, critical urban theory and urban sustainability. He has received research funding from the National Science Foundation and the International Olympic Committee. Before joining Clark University, Mark held fellowships at the Nelson A. Rockefeller Centre for Public Policy and Social Science, Dartmouth College, and the Urban Research Centre, University of Western Sydney. He holds a BA(Hons) and PhD in Geography from King's College London.

Oddrun Helen Hagen is a Master of Science in Physical Planning, and a Research Planner in the Sustainable Urban Development and Mobility group at The Institute of Transport Economics, Oslo. Before entering the research field 2 years ago, she worked for 13 years in planning practice. Her experiences cover sustainable urban planning and sustainable rating systems for master planning (BREEAM Communities), street design, large-scale infrastructure projects, public participation and planning processes. She has worked at all stages of planning, ranging from strategic planning to the construction stage. Ongoing research projects cover walkability, car-free city centres, changes in the transport systems, multimodality and access-egress.

Iqbal Hamiduddin is Lecturer in Transport Planning and Housing at the Bartlett School of Planning, University College London. He is particularly interested in the social impacts of different housing and transport regimes and the policy implications on well-being, social sustainability and future planning practice.

Meg Holden is professor of urban studies and geography at Simon Fraser University in Vancouver, Canada. Her research programme examines urban policy, planning and social aspects of sustainable development intentions and transitions in cities and communities. Her book *Pragmatic Justifications for the Sustainable City: Acting in the common place* was published in 2017 by Routledge. She is the principal investigator of a multi-year international research project that is investigating the policy and lived outcomes of initiatives to build model sustainable neighbourhoods worldwide.

Ramin Keivani is Professor of International Land Policy and Urban Development and Research Lead at the School of the Built Environment, Oxford Brookes University. He is an urbanist with a particular research interest on low-income and affordable housing policy, urban social sustainability, and impact of land and property markets on urban development and urban equity in both developed and transition economies. He has worked on projects across the globe including UK, central/eastern Europe, the Middle East, India, Brazil, China and Asia Pacific, and southern Africa. He is the founding and managing editor of the *International Journal of Urban Sustainable Development*.

Chris Landorf teaches architectural and heritage management and is a member of the Architecture Theory Criticism History research centre in the School of Architecture at the University of Queensland. She is a registered architect with postgraduate qualifications in business administration, facility management and the sustainable management of industrial heritage sites. Her key research areas are digital cultural heritage and the sustainability of historic industrial and urban environments. She recently co-convened conferences in Brisbane and London titled 'digital cultural heritage: FUTURE VISIONS', the latter in collaboration with University College London's Bartlett Real Estate Institute.

Richard Lang is Assistant Professor at the Institute of Innovation Management (IFI) at Johannes Kepler University Linz (JKU), Austria. Previously, he was a Research Fellow of the Austrian Academy of Sciences (APART

Programme) and Marie Curie Senior Research Fellow at the Housing and Communities Research Group at the University of Birmingham, UK. His research interests include cooperative and collaborative forms of housing, social innovation and social enterprise models and their role in urban and regional development.

Majken Toftager Larsen is a PhD student at the Planning Studies programme at Roskilde University in Roskilde, Denmark. In her PhD research she focuses on urban governance work and planners' methods for advancing strategies for sustainable urban development. As part of this she has been involved in the Interreg project Urban Transition Øresund which explored practices and experimented with new planning approaches in five cities in Denmark and Sweden. She is currently also involved in another European project investigating capacity building in local community organisations in order to meet challenges revitalising urban environments in residential towns.

Vinicius M. Netto holds a PhD in Advanced Architectural Studies (University College London), and a position as Associate Professor at the Department of Urbanism, Fluminense Federal University (UFF), Rio de Janeiro state. His work is centred on cities as networks of encounter, information and material interaction. He is the author of *The Social Fabric of Cities* (Routledge, 2017), and articles published in journals such as *International Journal of Urban and Regional Research* (IJURR), *Society and Space* (EPD) and *Complexity*. Vinicius is Editor for Latin America of *Area Development and Policy* (Regional Studies Association).

Renato T. de Saboya is Associate Professor of Architecture and Urbanism at the Federal University of Santa Catarina, Brazil, for both graduate and postgraduate levels. He received his Professional Degree in Architecture and Urban Design at UFSC (1997), a Masters' Degree in Urban and Regional Planning at Federal University of Rio Grande do Sul (UFRGS) (2001) and a Doctor's Degree in Civil Engineering at UFSC. His main research areas include urban and architectural morphology and configuration, accessibility, diversity, and segregation.

M. Reza Shirazi is Marie Skłodowska-Curie Global Fellow at the Institute of Urban and Regional Development (IURD), UC Berkeley, and Senior Research Fellow at the School of the Built Environment, Oxford Brookes University, UK. His research and teaching interests lie in the field of urban justice, discourse studies, neighbourhood development, citizen participation, phenomenology of the built environment, sociocultural sustainability, and architectural and urban transformation in the MENA (Middle East and North Africa) countries. Reza's recent books include: *Contemporary Architecture and Urbanism in Iran: Tradition, Modernity, and the Production of 'Space-in-Between'* (Springer, 2018), *Citizens' Participation in Urban Planning and Development in Iran; Challenges and Opportunities*

(Routledge, 2017), and *Towards an Articulated Phenomenological Interpretation of Architecture: Phenomenal Phenomenology* (Routledge, 2014).

Yamini Jain Singh is an urban and transport planner based in the Netherlands. Her research interests include sustainable mobility, smart mobility/cities and their impacts, geographic information system (GIS) and others. Over the last 14 years of her work experience, she has worked on various infrastructure projects funded by the World Bank and Asian Development Bank, published journal articles, given invited lectures at universities, advised PhD students and reviewed papers for multiple international journals. Her current projects relate to measuring bikeability, first and last mile connectivity and ride-sharing services. Her forthcoming book *Transit-oriented Development for Developing Countries* will be published by Springer.

Moa Tunström is Senior Research Fellow at Nordregio in Stockholm, Sweden. She is educated as a planner and holds a PhD in Human Geography from Örebro University (2009) in Sweden. Her research focuses on contemporary urban discourses and storytelling – in urban plans and visions, planning practice, education, etc. In projects and writings, she has investigated discourses on the urban and suburban in planning as well as social sustainability. She is currently involved in research on citizen participation in Swedish planning.

Tanu Priya Uteng is a senior researcher at the Institute of Transport Economics, Oslo. She has worked across a host of urban and transport planning issues in the past 16 years, and a few of her areas of expertise include: transport-related social exclusion, travel praxis and mobilities, transport modelling and gender studies. She is currently leading several long-term strategic research projects funded by the Norwegian Research Council looking at topics like bicycling, access–egress, climate change and travel behaviour, green-shift in the Nordic region, car-sharing and future mobilities. Her first edited book *Gendered Mobilities* was published in 2008 while her second edited book *Urban Mobilities in the Global South* came out in autumn 2017.

Júlio Celso Vargas is a licensed architect, holding a Masters' degree on Urban and Regional Planning and a Doctoral degree on Transportation Engineering from Universidade Federal do Rio Grande do Sul, granted with a visiting studies period at TRAC – Washington State Transportation Center – University of Washington, USA. He is a tenured professor at Federal University of Rio Grande do Sul's School of Architecture, member of the Urban Health, Environment and Inequalities Research Group and the NTU – Urban Technology Center. He is the author of several articles and book chapters, currently dedicated to the field of urban mobility, conducting research on non-motorised transport and urban morphology aided by geographic information system (GIS) and computational models.

1 Social sustainability discourse

A critical revisit

M. Reza Shirazi and Ramin Keivani

The 'social' in the sustainable development discourse: does social sustainability matter?

The idea of sustainability and sustainable development emerged during the late 1960s and early 1970s as a reaction to the growing environmental, economic, and social challenges worldwide. Two interconnected crises played a vital role in the emergence of sustainable urbanism: an ecological crisis as the result of the culmination of the environmental damages of rapid industrialization, and an urban crisis of deteriorating quality of urban life in the rapidly expanding cities worldwide (Whitehead, 2011). Although the roots of public awareness with regard to negative consequences of industrial development, urban growth, environmental degradation, social inequality, and economic injustice go back to decades before, two publications, *The Limits to Growth* (Meadows *et al.*, 1972) and *A Blueprint for Survival* (Goldsmith and Allen, 1972), suggested serious concerns about the future of our planet. The 1987 release of 'Our Common Future' report by the World Commission on Environment and Development, chaired by the Prime Minister of Norway Gro Harlem Brundtland, was a turning point: the concept of sustainable development became the mainstream globally (Wheeler, 2013). Several international meetings and conferences were organized around this concept, scholars of many fields published intensively to explore its advantages and challenges, city administrations produced plans and visions for a sustainable urban future, and national and local policy documents provided strategic planning guidelines to achieve sustainable urban and regional development.

The classic formulation of sustainable development that appeared in Brundtland's report defines sustainable development as a kind of 'development that meets the needs of the present without compromising the ability of future generations to meet their own needs' (World Commission on Environment and Development, 1987). It is out of the scope of this introduction to reflect critically on the scholastic debates on the definition, structure, validity, and contradictions of the concept of sustainable development. None the less, it is important for this collection to show the significance and relevance of 'the social' to the sustainable development discourse.

A repeatedly referred framework for sustainable development formulates it as the meeting point of 'three Es': economy, ecology, and (social) equity (Connelly, 2007; Opp and Saunders, 2013). This tripartite structure has, however, been subject to criticism from the beginning. To gain a clear image of the critiques provided, we can study them in two categories of reformist and revisionary approaches. The reformist approach accepts, or corresponds with, the tripartite structure of sustainable development formulation, namely three Es, but calls for efficient, intensive, and holistic balance and coordination between the Es in order to achieve greater sustainability (Neuman, 1998; Berke, 2002; Winston and Pareja Eastaway, 2008; Dale and Newman, 2010; Boyer *et al.*, 2016; Peterson, 2016). In this sense, sustainability is achieved only if there is a strong dialogue and interaction of three constituent elements of sustainable development. The revisionist approach, on the other hand, challenges the accuracy, comprehensiveness, and all-inclusivity of the tripartite structure, and suggests either a multi-pillar formulation or a totally new format (Hawkes, 2001; Godschalk, 2004; Duxbury and Jeannotte, 2010; Burford *et al.*, 2013; Soini and Birkeland, 2014; Leal Filho *et al.*, 2016). Reviewing some examples from both approaches shows the nature of their argumentation.

Reformist approach argues for a balance among three Es, because they are structurally interdependent, conceptually inseparable, and practically indistinguishable. For example, Peterson (2016) contends that sustainability should be understood as an integrated concept whereby social aspects are imbedded into environmental and economic aspects in an undistinguishable way. Campbell (1996) argues that the tripartite structure of sustainable development is a useful model for planning, but carries significant conflicting interests for three constituent components that need to be discursively examined and creatively reconciled. Boyer *et al.* (2016: 2) argue for an integrated, place-based, and process-oriented approach to sustainability that requires 'a framework that acknowledges the legitimacy of local knowledge and the importance of decision-making processes in policy and business'.

Revisionist approach normally argues for a 'missing pillar' (Burford *et al.*, 2013) in the sustainable development definition, and introduces it as an important aspect that has not been included in, and identified as an integral part of, the sustainability discourse. A number of concepts have been suggested as the missing pillar, such as culture, liveability, governance, politics, and ethical values. Culture has been repeatedly suggested as the most important missing pillar (Hawkes, 2001). Soini and Birkeland (2014: 214) argue that 'there have been no scientific studies systematically aimed at analysing and elaborating the role and meaning of culture in sustainable development and culture in the framework of sustainable development has remained under-emphasized and under-theorized'. They suggest 'cultural sustainability' as the fourth and parallel dimension to ecological, economic, and social sustainability, and obviously distinguishable from other dimensions:

> Cultural sustainability is linked but not equal to issues of social sustainability, such as social justice and equity, social infrastructure, participation and engaged governance, social cohesion, awareness, needs and work, and issues of the distribution of environmental 'goods' and 'bads'.
>
> (Soini and Birkeland, 2014: 214)

Other dimensions that have been suggested as supplementary pillar include 'governance' (Leal Filho *et al.*, 2016), 'political bottom line' (Bendell and Kearins, 2005), and 'ethical values' (Burford *et al.*, 2013), to name but a few.

Some revisionist approaches go further and propose a new format for sustainable development. For example, Godschalk (2004) proposes a 'sustainability/liveability prism' which is capable of guiding best practices in land use planning and discovering the interaction between the elements. Seghezzo (2009) suggests a five-dimensional sustainability triangle formed on three Ps: 'Place' which contains three dimensions of space, 'Permanence' which is the dimension of time, and 'Persons' as the fifth human dimension. Psarikidou and Szerszynski (2012) argue that the three-pillar structure leads to narrow, desocialized conceptions of nature and economy. They suggest a sociomaterial turn in the way we think about sustainability, which would be both social in the sense of attending to social relations, practices, cultural meanings, and normative judgements, and material in the sense that recognizes the conduction of social life by embodied beings in exchange with their physical environment. In this sense, 'the economic' is understood as embedded in social relations and 'the social' as including relations between humans and the material world, and thus boundaries between the three Es are dissolved. This approach, they argue, 'requires us to approach sustainability as a whole in a different way – as a lived, embodied form of life, with its own spatial organization and temporal rhythms' (Psarikidou and Szerszynski, 2012: 37).

This brief review confirms that, no matter how many dimensions or pillars sustainable development framework should have, and how these dimensions should interact and be integrated, the social aspect remains an integral component of any sustainable development framework (Shirazi and Keivani, 2017). In both reformist and revisionist approaches, the significance of the social dimension has not been challenged, but recognized and underlined to the extent that we could argue social sustainability does matter!

Social sustainability as an umbrella discourse

As is briefly noted in Chapter 6, studying 'the social' in the built environment is much older than the sustainable development debate; scholars have intensively and broadly investigated social dimension of cities and communities under a wide range of informative concepts. However, it is fair to argue that, by the advent of the sustainable development concept and its mainstreaming, many already existing and developed discourses with social concerns were assembled under the 'social sustainability discourse'. Here we would like to have a closer look at a number of similar concepts and explore areas of commonalities: social equity and justice, social capital, social cohesion, social exclusion, environmental justice, quality of life, and urban liveability.

Social equity and social justice has been one of the central concerns of planning since mid-twentieth century (Yiftachel and Hedgcock, 1993) as a basic human right (Magis and Shinn, 2009). It has both intra-generational and inter-generational aspects so that 'Both intra-generational equity providing for the needs of the least advantaged in society, and inter-generational equity, ensuring a fair treatment of future generations, need to be considered' (Elkin *et al.*, 1991: 203). Social equity and justice advocates the notion of distributive justice and fairness in the apportionment of resources in society, regardless of gender, race, ethnicity, and social status. According to Harvey (2009: 98), the principle of social justice starts with this skeleton concept: 'a just distribution justly arrived at'. Distributional justice can be seen from two perspectives: fairness of the outcome of distribution (the end result), and the fairness of the actions and procedures that generate the outcome (Burton, 2000: 1987). Social equity indicates equal access of services in terms of geography; all areas regardless of social or political character should have equal access to the services, what Kay (2005) refers to as horizontal equity. Social equity also addressees access to political and economic opportunities, and requires restructuring of power through wealth distribution, elimination of any type of socioeconomic and legal barriers, and removal of excessive political power from the minority (Magis and Shinn, 2009). Thus, social equity and justice provide an ethical ground for formulating social sustainability (Cuthill, 2010).

Social capital is

> the aggregate of the actual or potential resources which are linked to possession of a durable network of more or less institutionalized relationships of mutual acquaintance and recognition – or in other words, to membership in a group – which provides each of its members with the backing of the collectively-owned capital, a 'credential' which entitles them to credit, in the various sense of the word.
>
> (Bourdieu, 1986: 248)

Unlike physical capital, which is entirely tangible, social capital is less tangible because 'it exists in the *relations* among persons' (Coleman, 1988: 110–111). Three forms of social capital exist: obligations and expectations, information channels, and social norms. Obligations and expectations are based on trustworthiness and make a community powerful, social relations provide information that facilitates action, and social norms provide benefit for individuals, families, and groups within community. This term has been extensively used in the late twentieth century by academics, international institutions, and policy makers (World Bank, 1998; Forrest and Kearns, 2001; Grootaert and Bastelaer, 2001). Putnam played a significant role in further development of the concept. He defines social capital as the 'features of social organization, such as trust, norms and networks, that can improve the efficiency of society by facilitating coordinated actions' (Putnam *et al.*, 1993: 167). Social capital may have bonding (exclusive) qualities or bridging (inclusive) qualities; while bonding forms of social capital are inward looking and 'reinforce exclusive identities and homogenous groups', bridging qualities are 'outward looking and encompass people across diverse social cleavages' (Putnam, 2000: 22). EU and UK urban policy has been influenced by Putnam's interpretation of the concept of social capital (Manzi *et al.*, 2010).

Social cohesion has been an enduring subject of inquiry for sociologists and psychologists (Friedkin, 2004). It refers to the force that holds individuals within a group, and is associated with social interaction, social networks, sense of belonging, and community engagement (Raman, 2010). Social cohesion emphasizes

> the need for a shared sense of morality and common purpose; aspects of social control and social order; the threat to social solidarity of income and wealth inequalities between people, groups and places; the level of social interaction within communities or families; and a sense of belonging to place.
>
> (Forrest and Kearns, 2001: 2128)

In the contemporary cities, high crime rate, unemployment, organized crime, and low living standard, all point to weak social cohesion in the cities. This generates social disorder and conflict, disparate moral values, social injustice, social exclusion, and place dis-attachment.

Social exclusion was first used in France in the 1970s to describe the condition of certain marginalized groups cut from the benefits of the welfare state, though it later gained a wide currency in the social policy Europe-wide and across the globe (Pierson, 2010). It refers to a 'process over time that deprives individuals and families, groups and neighbourhoods of the resources required for participation in the social, economic and political activity of society as a whole' (Pierson, 2010: 12). An individual is socially excluded 'if he or she does not participate in key activities of the society in which he or she lives' (Burchardt *et al.*, 2002a: 30).

Social exclusion is a slippery and contested concept (Gough and Eisenschitz, 2006; Taket *et al.*, 2009) and covers a wide range of issues such as unemployment, low income, low educational attainment (lack of skills), bad health, poor housing, and high crime (Social Exclusion Unit, 1998). Although poverty has been widely considered as a main reason for social exclusion, it is a phenomenon distinct from poverty and also economic inequality because social exclusion can happen between groups that are not suffering from poverty (Barry, 2002). Moreover, individuals may suffer from the above-mentioned problems but still be integrated and part of a mainstream society. Another significant issue is the political exclusion, which indicates exclusion from access to decision-making processes (Davoudi and Atkinson, 1999) and not participating in many of the normal activities of the society (Turok *et al.*, 1999). Different views exist with regard to fundamental causes of social exclusion, such as individuals' behaviour and moral values, the role of institutions and systems, and issues of discrimination and lack of enforced rights (Burchardt *et al.*, 2002b). The 'place' of living is a significant factor in social exclusion, because people and communities are excluded by virtue of their living location, in particular urban areas. This spatial manifestation of social exclusion is concentrated in deprived inner or peripheral urban areas, and 'is constructed through the physical organization of space as well as through the social control of space, as ensured by informal codes and signs and formal rules and regulations' (Madanipour, 1998: 86–87).

Environmental justice advocates the idea that public actions should not disproportionately disadvantage any social group (Agyeman and B. Evans, 2004; Agyeman and T. Evans, 2003), and environmental problems should not 'bear down disproportionately upon the poor' (Agyeman *et al.*, 2003: 1). It deals with the question of distribution: who, in terms of gender, race, and class, gets negative impacts of environmental 'bads' and who benefits from the environmental 'goods' (Boström, 2012). It also concerns any procedures that result in distributional inequities (Faber and McCarthy, 2003). Agyeman coined the term 'just sustainability' to link sustainable development to environmental justice: promoting 'an equal concern with equity, justice and, ultimately, governance on the one hand, and environment on the other' (Agyeman and B. Evans, 2004: 160).

Quality of life as an informative idea has attracted scholar's attention since the 1930s (Massam, 2002), though it was around 1960 that this concept as a research field came into being (Schuessler and Fisher, 1985). Broadly speaking, it refers to the good and satisfactory character of people's lives (Szalai, 1980), the ability to adapt a lifestyle that satisfied the individual's needs and wants (Karen *et al.*, 1990), and 'the individual's achievement of a satisfactory social situation within the limits of perceived physical capacity' (Lamendola and Pellegrini, 1979: 457). The World Health Organization (WHO) defines quality of life as 'individuals' perceptions of their position in life in the context of the culture and value systems in which they live and in relation to their goals, expectations, standards and concerns' (WHO, 1995: 1). It appears as an umbrella term and covers a wide range of fields and concepts such as education, physical and mental health, economic prosperity, happiness, liberty, standard of living, participation, environmental quality, urban services, and housing. Some scholars define this concept so broadly that it encompasses almost every aspect of urban life including environmental, social, physical, political, psychological, and economic (Serag El Din *et al.*, 2013). However, due to the value-laden character and subjective nature of the concept, a universally accepted definition is hard to achieve (Romney *et al.*, 1994). Three critical attributes can define the concept: individuals make a subjective appraisal of their own lives; they identify their satisfaction with life as it pertains to the physical, psychological, and social domains of their life; and finally objective measures may supplement their subjective evaluations of the quality of life (Mandzuk and McMillan, 2005).

Quality of life is dependent on exogenous (objective) facts of individuals' lives, as well as their endogenous (subjective) perception of these factors (Dissart and Deller, 2000), and thus encompasses both environmental and psychological aspects. Although the objective dimension addresses concrete aspects of the built environment, the subjective dimension deals with the individual's satisfaction with certain aspects of life. However, recent debates underline that quality of life is a purely subjective experience, because it is determined by one's subjective appraisal of the life condition (Moons *et al.*, 2006). This subjectivity implies that quality of life is not static, but may alter during time negatively or positively. It is assumed that quality of life can be controlled and modified. This is why it has always been a concern for planners, politicians, decision-makers, non-governmental bodies, and the public, to influence planning processes and outcomes in such a way as to protect and enhance the material and sociocultural milieu to secure improvement of quality-of-life indicators (Dennis *et al.*, 1993; Massam, 2002; Soleimani *et al.*, 2014).

Urban liveability has been a dominant category of urban discourse since the 1960s (Kaal, 2011). It addresses a set of environmental characteristics that make a place attractive to live, work, and invest in. Liveability 'relates to those attributes of a place, ranging in scale from dwelling and neighbourhood to a city and its region that contribute to residents' quality of life and well-being' (Newton, 2012: 82), and encompasses 'those elements of home, neighborhood, and metropolitan area that contribute to safety, economic opportunities and welfare, health, convenience, mobility, and recreation' (Vuchic, 1999: 7). Liveability in urban context goes beyond physical characteristics and as an inherently anthropocentric concept reflects quality of life, well-being, and satisfaction of the people with their needs (de Haan *et al.*, 2014), and thus includes the social dimension of urban life:

> Liveable cities are socially inclusive, affordable, accessible, healthy, safe and resilient to the impacts of climate change. They have attractive built and natural environments. Liveable cities provide choice and opportunity for people to live their lives, and raise their families to their fullest potential.
>
> (Department of Infrastructure and Transport, 2011)

Definitions of liveability, and consequently relevant indicators, vary in different geographies depending on place and local values (Southworth, 2003).

Overall, liveability has both objective and subjective dimensions and indicates the degree of satisfaction of the residents with objective and subjective dimensions of the urban environment, what is referred to as perceived liveability (Namazi-Rad *et al.*, 2016). Urban liveability is also substantially correlated with urban sustainability in that it has a profound focus on needs and an ability to satisfy them (de Haan *et al.*, 2014). Scholars warn about a narrow definition of liveability to serve wishes and needs of the mobile professional élites at the expense of the lives of ordinary city residents, and call for reorienting the term towards the realization of more inclusive and socially progressive policy agendas (Bunnell and Kathiravelu, 2016).

This brief review provides us with enough evidence to suggest that social aspects of the built environment and societies have been subject to scientific scrutiny long before sustainable development became mainstream in the 1990s. In this sense, social sustainability, as an integral dimension of urban sustainability, benefitted from previously developed and examined disciplines, utilized their achievements and arguments, and gathered them under an umbrella. Social sustainability advocates just distribution of welfare goods and urban services (Murphy, 2012), fosters enriching social capital and maintaining it at sustainable levels for future generations (Ročak *et al.*, 2016; Dillard *et al.*, 2009), promotes stronger social cohesion and community ties (Forrest and Kearns, 1999), considers social exclusion as a danger for sustainable communities (Polèse and Stren, 2000), addresses environmental justice through equitable re-distribution of environmental 'goods' and 'bads' (Agyeman *et al.*, 2002), and envisions improving the quality of life of the people now and in the future (Woodcraft, 2012; Weingaertner and Moberg, 2014).

The challenge of social sustainability discourse

Scholars have criticized the concept of social sustainability from different perspectives. According to Manzi *et al.* (2010: 20), the concept of social sustainability faces four main criticisms: it is too abstract to be practically implemented; it fails to address the complexity of local political contexts; it does not acknowledge the basic constraints of empowerment and participation agenda; and only lip service is paid to the international and global dimension. Boström (2012: 12) lists six challenges for operationalizing and integrating social sustainability: high expectation; vague, subjective, and ideological framing; historical roots (sustainability framing is better suited to environmental than social issues); missing institutional linkage; global capitalism for sustainable development; and the relationship between the procedural and substantive dimensions of social sustainability. According to Boyer *et al.* (2016: 1–2) the current opacity around social sustainability 'results from a multiplicity of legitimate meanings, lack of cross-disciplinary communication, and a reluctance to engage diverse and local sources of knowledge in scholarly research'.

The concept of social sustainability, we argue, faces critical challenges in areas of theory and operation. In what follows, we discuss these areas of challenge in detail.

Theoretical challenges of social sustainability

Overall one can argue that the social sustainability discourse is under-theorized, ambiguous in definition, limited in disciplinary coverage, geographically concentrated on a limited number of developed countries, and normative in character.

Under-theorized discourse

Sustainability has been introduced as 'a topic of research that is basically social' (Becker and Jahn, 1999: 4). However, although ecological and economic pillars of sustainable development have been widely debated, discussed, and developed, social pillar has remained under-theorized, so that social sustainability is still in its formative stages within the sustainability dialogue (Magis and Shinn, 2009). In fact, as Dillard *et al.* (2009: 2) put it: 'concerns with environmental and economic sustainability have eclipsed efforts to understand the social aspects of sustainability'. Consequently, scholars have repeatedly argued that the concept of social sustainability has not been well developed (Yiftachel and Hedgcock, 1993; Koning, 2002; Bramley *et al.*, 2009; Cuthill, 2010; Colantonio and Dixon, 2011; Woodcraft, 2012; Ročak *et al.*, 2016; Yoo and Lee, 2016), to the extent that a clear theoretical concept of social sustainability is missing (Littig and Griessler, 2005). This is partly due to the less tangible nature of social consequences of an unsustainable development: we rightly highlight the environmental disadvantages of air pollution, flooding, and global warming, but reluctantly make social disadvantages of urban inequalities, insecurity, and poverty visible. Overall, despite recent attention to social dimensions, social sustainability remains relatively under-theorized and poorly discussed.

Ambiguity of definition

Social sustainability is 'a concept in chaos' (Vallance *et al.*, 2011: 342). There is no consensus on the definition of this concept (Dillard *et al.*, 2009; Weingaertner and Moberg, 2014); a 'single blueprint definition' (Weingaertner and Moberg, 2014: 123) does not exist. Individuals derive their own understanding according to discipline-specific criteria of particular study perspective (Colantonio and Dixon, 2011), and thus generate a plethora of definitions. Table 1.1 provides a non-exhaustive overview to the existing definitions of social sustainability in the literature.

A closer look at Table 1.1 suggests that diversity of definitions with regard to social sustainability is identified by the way in which this concept is approached. In other words, factors underpinning our view towards social sustainability determine the way we define it. Overall, five factors of scale, time, discipline, key themes, and population are the main determinants of the definitions. Scale is the geographical setting, from neighbourhood to city and

Table 1.1 Definitions of social sustainability

'The continuing ability of a city to function as a long-term viable setting for human interaction, communication and cultural development.... Urban social sustainability is about the *long-term survival* of a viable urban social unit' (Yiftachel and Hedgcock, 1993: 140) [original emphasis].
A strong definition of social sustainability must rest on the basic values of equity and democracy, and include all human rights such as political, civil, economic, social and cultural for all the people (Sachs, 1999).
'Social sustainability for a city is defined as *development (and/or growth) that is compatible with the harmonious evolution of civil society, fostering an environment conducive to the compatible cohabitation of culturally and socially diverse groups while at the same time encouraging social integration, with improvements in the quality of life for all segments of the population*' (Stren and Polese, 2000: 15–16) [original emphasis].
Social sustainability 'refers to a society that is socially just, equal, without social exclusion and with a decent quality of life, or livelihood, for all' (Koning, 2002: 70).
Social sustainability is 'a life-enhancing condition within communities, and a process within communities that can achieve that condition' (McKenzie, 2004: 12).
Socially sustainable urban transportation 'is understood as transportation that provides equitable access to urban opportunities, minimizes social exclusion, and improves or does not overly diminish an individual's quality of life' (Boschmann and Kwan, 2008: 139).
'Social sustainability concerns the ability of human beings of every generation to not merely survive, but to strive' (Magis and Shinn, 2009: 38).
Social sustainability refers to 'The processes that generate social health and well-being now and in the future' (Dillard *et al.*, 2009: 4)
Social sustainability is 'a positive condition within societies that support human well-being and a process within communities that can achieve that condition' (Dujon, 2009: 122).
Social sustainability is 'a practice of maintenance, the establishment of social arrangements that enable democratic politics to remain 'in balance''' (Davidson, 2009: 615).
The concept of social sustainability addresses both 'social equity' (access to services, facilities and opportunities) and the 'sustainability of community itself' (the ability of society to sustain and reproduce itself at an acceptable level of functioning) (Bramley *et al.*, 2010).
Three types of social sustainability include: 'development sustainability' which addresses poverty and inequity, 'bridge sustainability' which concerns changes in behaviour so as to achieve biophysical environmental goals, and 'maintenance sustainability' which addresses preservation of sociocultural patterns and practices in in the context of social and economic change (Vallance *et al.*, 2011).

Table 1.1 Continued

'Social sustainability concerns how individuals, communities and societies live with each other and set out to achieve the objectives of the development models that they have chosen for themselves, also taking into account the physical boundaries of their places and planet earth as a whole' (Colantonio and Dixon, 2011: 24).
'Social sustainability is about people's quality of life, now and in the future. It describes the extent to which a neighbourhood supports individual and collective well-being' (Bacon *et al.*, 2012: 9).
'With social sustainability, we understand the output of local relations, interests and capacities promoting the daily well-being of the people in the neighbourhood and promoting the ability to act collectively when needed' (Søholt *et al.*, 2012: 256).
'For a city to be labelled as socially sustainable, all people, regardless of race, ethnicity, gender, or income level must, have the ability to enjoy equal access to the fruits of public investment while also being able to satisfy their basic needs' (Opp, 2016: 6).
Social sustainability in an urban shrinkage context is defined 'as a resilient outcome of the interaction of civil society and other stakeholders in society enabling quality of life and the social transformation of shrinking cities' (Ročak *et al.*, 2016: 1).
'Social sustainability is defined as the ability to maintain one society's certain state or existence' (Yoo and Lee, 2016: 6).
'Social sustainability comprises socially oriented practices intended to address major social issues to cope with the risks of climate change and environmental hazards.... Social sustainability strives to confront risk while addressing social concerns' (Eizenberg and Jabareen, 2017).
'Socially sustainable neighbourhoods are localities where conceived and perceived qualities are highly valued and interactively practiced by the inhabitants for a considerable period of time' (Shirazi and Keivani, 2018).

beyond; time is the temporal coverage and the way in which past, present, and future are mingled and intertwined; discipline is the disciplinary arena within which social sustainability is debated, e.g. transportation, urban regeneration, housing; key themes refer to the key societal focus of the approach, which could be a single one or a combination of some; and population is the number of people addressed, from individual through family to larger community. The way these factors are determined thus sets the ground for the way social sustainability is defined. And this diversifies the resulting definitions.

Although some scholars advocate constructing an acceptable and comprehensive definition for social sustainability in each discipline, e.g. transportation (Boschmann and Kwan, 2008), others evaluate the 'elastic and ever-moving' concept of social sustainability as positive, because users can shape it to their own circumstances (Manzi *et al.*, 2010: 21). In this sense,

'The lack or a coherent definition of social sustainability is not, therefore, something that should be derided or bemoaned, but rather accepted as a natural part of the sustainability agenda' (McKenzie, 2004: 21). In other words, as Davidson (2010: 877) puts it: 'The diversity of approaches to social sustainability is both inevitable and desirable.' Diversified approaches and definitions enrich the social sustainability discourse, push the borders of knowledge, and encourage dialogue with other discourses.

Disciplinary coverage

It has been argued that social sustainability is an essentially cross-disciplinary concept, covering a broad range of knowledge from natural science to social science and humanities (Åhman, 2013). According to Dillard *et al.* (2009), in order to adequately address multidimensionality of social sustainability, different expertise from a wide range of disciplines should come together, including anthropology, sociology and cultural studies, public administration, political science, social work, public health, architecture, environmental studies, business, economics, etc. Despite the fact that, as noted, social sustainability is a less-developed dimension of sustainable development and the number of researches available are limited, existing literature suggests that it has been examined in a wide variety of fields and disciplines: density and social sustainability (Bramley *et al.*, 2010; Dave, 2010, 2011; Dempsey *et al.*, 2012); neighbourhood environment and social sustainability (Karuppannan and Sivam, 2011; Yoo and Lee, 2016); urban form and social sustainability (Bramley and Power, 2009; Bramley *et al.*, 2009); social sustainability, regeneration, and urban renewal (Chan and Lee, 2007; Colantonio and Dixon, 2011; Križnik, 2018); social sustainability of housing provision (Chiu, 2002, 2003; Ancell and Thompson-Fawcett, 2008); social sustainability and well-being (Kytta *et al.*, 2016); social sustainability of urban interventions (Søholt *et al.*, 2012); evaluation of social sustainability of heritage sites (Landorf, 2011; Yung and Chan, 2012); social sustainability of shrinking cities (Ročak *et al.*, 2016); social sustainability of community gardens (Rogge *et al.*, 2018); social sustainability of mega sport events (Smith, 2009, 2010; Fleischer *et al.*, 2013); social sustainability, mobility, and transport (Boschmann and Kwan, 2008; Grieco, 2015); social sustainability and ageing (Hamiduddin, 2015); social sustainability and urban foodscape (Psarikidou and Szerszynski, 2012); and social sustainability and climate-resilient communities (Baldwin and King, 2018). This list looks exhaustive and very broad in coverage; however, we would argue, there are a large number of disciplines and fields that either have not been subject to scrutiny in social sustainability discourse, or have been inadequately touched. As an example, gender equity is one of the areas that has been rarely investigated in the social sustainability discourse: 'Social sustainability and gender, in particular, is still too infrequently 'on the radar', with consequence that we have not yet directly addresses the issue of systematic gender representation' (Turner, 2012: 52). This is why, we would argue, future social sustainability research should move forward in two disciplinary directions: deepening knowledge depth in 'a' particular discipline, and broadening knowledge scope to include new disciplines.

Geographical coverage

Social sustainability reflects its particular context and needs to be studied in a context-sensitive manner (Kytta *et al.*, 2016). Context implies the geographical realm in which social sustainability is discussed. Overall, social sustainability research has been largely limited to developed countries, or a small number of non-western countries mainly from the developing countries of south Asia and Latin America (Burgess, 2000; Karuppannan and Sivam, 2011; Cilliers, 2018), and thus has acquired a European, western-inflected interpretation. As social sustainability is rooted in sociocultural values, any generalization of research findings will be problematic (Shirazi and Keivani, 2017). What is urgent is broadening the geographical coverage of social sustainability research to include different types of countries, cities, and societies, to achieve place-specific observations and evidence, on the one hand, and enable cross-cultural, inter-regional comparisons, on the other, and thus gain a broader and more detailed understanding from its specifications and commonalities.

Normative character

Social sustainability discourse is normative in character. This normativity derives from the very normative nature of planning discipline. Planning is an inherently normative enterprise (Næss, 1994; Stein and Harper, 2005) and this urges planners to search for a 'better' future, think of what 'better' could and should mean (Healey, 2012), and address the very question of what should be done (Campbell, 2012; Winkler and Duminy, 2016). To formulate the 'better' and inform what should be done, planning defines a set of criteria informed by normative values, ethics, and ideas, prioritizes some ethical judgements over others, imagines the future, and recommends pathways to achieve this imagined condition. Much of the social sustainability discourse inherits the normativity of planning discipline, to the extent that indicators of social sustainability, as discussed later, address normative values and intend to respond to the question 'what should be done'. Questioning the normative nature of social sustainability discourse, we would argue, means opening up new perspectives to the debate, advocating innovative paths of theorization and operationalization that is context situated, and thus going beyond universal norms to localize and contextualize 'the better'.

Operational challenges of social sustainability

Scholars have highlighted the scarcity of social sustainability evaluation methods and the lack of sufficient measurement frameworks (Landorf, 2011;

Shirazi and Keivani, 2018). Operationalizing social sustainability and developing working frameworks for evaluation purposes need initial clarification with regard to the number and nature of indicators, the scale of inquiry, the geography of operation, and the field of operation.

Number and nature of indicators

A main challenge of operationalizing social sustainability is identifying the most relevant and appropriate indicators. This includes both the number and the nature (qualitative or quantitative) of indicators: what the most relevant indicators are, and which character they have. Identifying the most relevant indicators, of course, requires preliminary but broad insight into the discipline within which we are working, including an extensive knowledge of the previous and on-going debates on the field, followed by a proper research methodology that secures the identification of right and appropriate themes. The nature of indicators reflects their qualitative or quantitative character. Some scholars have observed a shift in the nature of social sustainability indicators from tangible (hard) indicators such as equity and poverty reduction to mainly intangible (soft) indicators such as identity and sense of place (Colantonio, 2009). This shift unravels the underlining qualitative essence of social sustainability and necessitates moving towards mixed methodologies in order to comprehensively address the complex and dynamic nature of social sustainability themes. The need for qualitative input into social sustainability measurement has already been underlined by some scholars (Chiu, 2003; Boschmann and Kwan, 2008). There is also a measurement challenge for operationalizing social sustainability indicators: although quantitative indicators are easy to assess, they are less informative than qualitative indicators, which are more complicated to measure. In many cases, mixed indicators and methodology would be more successful in achieving more fine-tuned understanding from the social sustainability dimension.

Scale of operation

The scale of operation is a key factor in developing evaluation frameworks, as the appropriateness of indicators has a scale and size dimension that must be considered. This implies that an indicator appropriate for the local scale may lose its significance for the regional scale, and vice versa. Thus, indicators should be identified depending on the scale of inquiry, from nation to community (Yoo and Lee, 2016). Not all the geographical scales have been equally subject to scientific scrutiny from the point of view of social sustainability. It seems that city scale has been the most studied scale and, as Hamiduddin (2015) puts it, little attention has been given to the social sustainability at the neighbourhood scale. The growing significance of urban neighbourhoods in different areas of urban planning and design suggests social sustainability of urban neighbourhoods as an area of concern for both researchers and policymakers (Shirazi and Keivani, 2018). The proper scale for operationalizing is another challenge, and may differ from discipline to discipline and country to country. For example, Boschmann and Kwan (2008) argue that in the United States urban regions provide an ideal context for dealing with socially sustainable urban transportation systems.

Geography of operation

Social challenges of urban settlements differ from city to city, and more importantly from country to country. As Grieco (2015: 87) puts it: 'The variations in the specificities of urban localities are likely to work against the adaptation of any one common practice or measurement or assessment.' Cities of the developing world have different social challenges to those of the developed world. Although poverty may stand at the top of the list for a city from a developing country, social cohesion and inclusion could be considered as the first priority in another city. This implies that indicators of social sustainability should be specified based on socioeconomic realities of the cities. To address this challenge and take into account the particularity of place, Landorf (2011), in her attempt to develop a framework for evaluating social sustainability in historic urban environments, suggests only three concepts of social equity, social coherence, and basic needs as principle dimensions, but lists a number of possible criteria for each indicator, which should be tailored for the given environment through a three-stage process: first, negotiating a definition of sustainable development at the local level; second developing a collective vision; and finally determining relevant indicators for evaluation. In this sense, the 'place' of operationalization matters, and place covers a wide range of geographies from neighbourhood through city to country and beyond.

Field of operation

Finally, social sustainability indicators are discipline specific, and should be tailored to match disciplinary specifications. Although some indicators such as equity could be found appropriate for most, if not all, of the disciplines, the meaning and implication of equity differ from discipline to discipline, and this requires a discipline-oriented definition of social sustainability indicators.

Contributing to the debate

This book contributes to the ongoing social sustainability discourse in the built environment and addresses the above-mentioned theoretical and operational concerns and challenges. The book provides a multidisciplinary understanding of the social sustainability of the built environment, because it covers a wide range of topics such as transportation, urban policy, neighbourhood development, urban heritage, and housing. It also brings together case studies from across the world, and thus offers new insights into the place specificity of social sustainability. Diverse definitions of social sustainability are introduced and debated, which underlines the potential of the concept for developing subject-specific interpretations. All this contributes to progressing social sustainability theory some steps further and paves the way for future developments. Many of the chapters are case study based, from different geographies, and have suggested clear methodologies to evaluate success and failure of the initiatives, programmes, policies, etc. They highlight challenges and potentials with regard to operationalizing social sustainability in different disciplines and application of measurement frameworks for evaluation purposes. Finally, most of the chapters provide us with practical recommendations and signpost pathways towards integrating social sustainability into urban planning policies and governance systems through participatory and collaborative mechanism.

In Chapter 2, Mark Davidson explores the political meaning of social sustainability. After underlining normative nature of social sustainability he suggests that it operates as an 'empty signifier': devoid of an inherent content but with an organizing duty. In this sense, social sustainability turns disordered thoughts into a coherent understanding. However, such an approach to social sustainability is not without contradiction and political difficulty: it is both indispensable and inherently problematic. Davidson argues that, similar to many social concepts, social sustainability as an empty signifier should be critically examined to ascertain how it can serve as a useful construct in any one time. His suggestion is that social sustainability must be supplemented with a commitment to democracy and equality, and establish a dialogue with democratic politics.

Moa Tunström, in Chapter 3, argues that contemporary social sustainability discourse in the Nordic cities must be understood through its relationship to the Nordic welfare state. Despite the socially oriented character of planning in the welfare system, reflected, for example, in planning and development of housing and neighbourhoods, Nordic cities face challenges in areas of socioeconomic inequality, urban segregation, reduction in welfare provision, and immigration. Tunström argues that the recent policy shift towards increased privatization and marketization of previously public services has weakened the legacies of the Nordic welfare state. On the other hand, rooted in the active civil society tradition, social sustainability in the Nordics is primarily a government project influenced by the welfare state, rather than a grassroots initiative. She reviews how three concepts of 'local democracy', 'neighbourhood', and 'mixed housing' in Nordic cities serve as tools for promoting social sustainability.

Chapter 4 asks how social sustainability could be understood and implemented in the transportation sector. Tanu Priya Uteng, Yamini Jain Singh, and Oddrun Helen Hagen discuss the extent to which smart mobility is socially sustainable, and explore how social sustainability in the transport sector can be inbuilt within the smart cities agenda from the inception of the projects. They evaluate accessibility, safety, health, cohesion, and awareness (information availability) of smart mobility, referring to different case studies and existing surveys. They conclude that transport needs to be decoupled from a purely technocratic assessment and fused with evaluations on social sustainability. Any transportation project should, therefore, entail social sustainability criteria, such as affordability, availability, and safety, right from the beginning.

In Chapter 5, Chris Landorf analyses four World Heritage urban landscape nomination documents and management plans to explore to what extent they have integrated four social sustainability dimensions: situation analysis, strategic orientation, stakeholder values, and stakeholder participation. The focus is on historic urban cultural landscapes such as dynamic cultural settings with multiple overlapping identities. The analysis shows that all four cases fail to satisfactorily integrate social sustainability principles into the heritage planning and management process, although two of them presented some indications to do so. Landorf concludes that mandatory requirements should be developed for socially sustainable development within World Heritage properties, and strategic planning needs to be more holistic and inclusive, so that sustainable development objectives are actively amalgamated into planning, implementation, and evaluation processes.

In Chapter 6, M. Reza Shirazi and Ramin Keivani argue that social sustainability is a sociospatial construct: it simultaneously and equally underlines the 'spatiality of place' and the 'sociality of space'. In other words, spatiality of social sustainability is constructed through objective spatiality of physical objects, as well as the intersubjective spatiality of individuals' and urbanites' perception. Although the former is essentially about the theory of urban form, the latter is related to the social theory of space. This chapter concentrates on the objective spatiality of social sustainability, suggests a methodology to map outdoor social activities in urban neighbourhoods, and applies it to the case of the Bethnal Green neighbourhood in London. The analysis shows significant correlation between characteristics of urban form and outdoor social activities in terms of type of activities, and gender and age of actors. Measuring spatiality of social sustainability provides policy makers, planners, and designers with necessary knowledge about the spatial pattern of social activities in order to promote sociality of urban space.

Vinicius M. Netto, Júlio Celso Vargas, and Renato T. de Saboya in Chapter 7 ask how architecture can create socially vibrant public spaces and cities? They investigate material condition of social sustainability, the co-presence of people in public spaces, and propose a method to identify elementary factors in the relationship between the built environment and social sustainability, and apply it to 46 randomly selected areas in 3 major capitals in Brazil. They argue that architectural fabric in larger cities in Brazil has been eroded coincidental with a progressive erosion of the fabric of co-presence in their public spaces. The chapter finds consistent relationships within complexities and contingencies in the connections of the 'social city' to the 'physical city', and that the architectural features close to the bodily appropriation of space are an active part of this relationship. The predominant isolated architectural types in major Brazilian cities break the relationship between built form and open forms of social life, and this endangers social sustainability of the environment. This calls for the establishment of stronger institutional frameworks, practical design guidelines, and performance-evaluation planning systems.

Meg Holden, Annika Airas, and Majken Toftager Larsen, in Chapter 8, show how the idea of social sustainability, despite different interpretations, shares many characteristics with the Nordic welfare state model, such as high standard of living for women, and relatively low inequality and poverty. But this creates a challenge for Nordic cities: the conflict of values that rises as the result of exchange and dialogue between internationally developed, socially sustainable values and already existing welfare system landscapes and habits. This is examined in some Nordic eco-urban neighbourhoods in Sweden, Finland, and Denmark. They argue that practices of social sustainability being generated within these neighbourhoods do not strongly follow the ideals that made the Nordic model successful in the first place; instead they generate exclusions and departures from long-standing social values.

Chapter 9 provides a critical examination of the implementation of social sustainability principles into the design and development of two new neighbourhood developments in Germany and Spain. Iqbal Hamiduddin and Marco Adelfio suggest a framework for evaluating social sustainability of neighbourhoods based on the two factors of social equity and community sustainability. The analysis shows that these neighbourhoods, although different in design and planning details, have been successful in addressing many aspects of social equity and community sustainability through attracting a range of resident groups, providing a basic 'safety net' of amenities and services within the neighbourhood, and good public transport access to their respective wider urban area. The chapter concludes with two recommendations: new neighbourhoods should include a diversity of land uses, with provision for a high level of local employment, and create diverse and inclusive spaces that nourish social relationships between residents, and reaffirm a sense of belonging.

In Chapter 10, Richard Lang critically analyses the contribution of collaborative housing models to sustainable communities and cities. Lang concentrates on two particular aspects of social sustainability – social inclusion and social cohesion – and draws on social capital as a suitable analytical perspective. Case studies are six collaborative housing projects in England and Austria. The analysis suggests that the creation of bonding social capital – community cohesion – represents the core of collaborative housing projects. Collaborative housing models benefit from a range of vertical linking strategies that connect resident communities with powerful external stakeholders. Moreover, collaborative models developed within subsidized housing programmes are successful in addressing the cohesion and inclusion aspects of social sustainability. This means affordability for different sociodemographic groups, but needs sustained political support.

In the concluding chapter, M. Reza Shirazi and Ramin Keivani first explain how different chapters of the book contribute to theoretical and operational challenges of social sustainability discourse outlined in this chapter. They then explore policy challenges of urban social sustainability and argue that to tackle them urban social sustainability should be recognized as an integral part of urban planning and policy. Moreover, transformative methodologies should be developed to incorporate social sustainability concerns into urban development planning using collaborative and dialogic procedures of decision-making that seek for inclusive and just urban interventions and projects. This could be achieved by means of establishing a functioning governance system for urban social sustainability. The chapter concludes that urban social sustainability should be treated as an urban agenda that recognizes its significance and governs its planning, implementation, and monitoring. The authors suggest a framework for such an approach and explain its main features and characteristics.

References

Agyeman J and Evans B (2004) 'Just sustainability': The emerging discourse of environmental justice in Britain? *Geographical Journal* 170(2): 155–164.

Agyeman J and Evans T (2003) Toward just sustainability in urban communities: Building equity rights with sustainable solutions. *Annals of the American Academy of Political and Social Science* 590(1): 35–53.

Agyeman J, Bullard RD, and Evans B (2002) Exploring the Nexus: Bringing together sustainability, environmental justice and equity. *Space and Polity* 6(1): 77–90.

Agyeman J, Bullard RD, and Evans B (eds) (2003) *Just Sustainabilities: Development in an Unequal World*. Cambridge, MA: The MIT Press.

Ancell S and Thompson-Fawcett M (2008) The social sustainability of medium density housing: A conceptual model and Christchurch case study. *Housing Studies* 23(3): 423–442.

Bacon N, Cochrane D, and Woodcraft S (2012) *Creating Strong Communities. How to Measure the Social Sustainability of New Housing Developments*. London: The Berkeley Group.

Baldwin C and King R (2018) *Social Sustainability, Climate Resilience and Community-Based Urban Development: What about the People?* New York: Routledge.

Barry B (2002) Introduction. In: Hills J, Grand JL, and Piachaud D (eds), *Understanding Social Exclusion*. Oxford, New York: Oxford University Press, pp. 13–29.

Becker E and Jahn T (1999) Fostering transdisciplinary research into sustainability in an age of globalization: A short political epilogue. In: *Sustainability and the Social Sciences: A Cross-Disciplinary Approach to Integrating Environmental Considerations into Theoretical Reorientation*. London, New York: New York: Zed Books, pp. 1–22.

Bendell J and Kearins K (2005) The political bottom line: The emerging dimension to corporate responsibility for sustainable development. *Business Strategy and the Environment* 14(6): 372–383.

Berke PR (2002) Does sustainable development offer a new direction for planning? Challenges for the twenty-first century. *Journal of Planning Literature* 17(1): 21–36.

Boschmann EE and Kwan M-P (2008) Toward socially sustainable urban transportation: Progress and potentials. *International Journal of Sustainable Transportation* 2(3): 138–157.

Boström M (2012) A missing pillar? Challenges in theorizing and practicing social sustainability: introduction to the special issue. *Sustainability: Science, Practice, & Policy* 8(1): 3–14.

Bourdieu P (1986) The forms of capital. In: *Handbook of Theory and Research for the Sociology of Education*. New York: Greenwood, pp. 241–258.

Boyer R, Peterson N, Arora P, *et al.* (2016) Five approaches to social sustainability and an integrated way forward. *Sustainability* 8(9): 878.

Bramley G and Power S (2009) Urban form and social sustainability: The role of density and housing type. *Environment and Planning B: Planning and Design* 36(1): 30–48.

Bramley G, Dempsey N, Power S, *et al.* (2009) Social sustainability and urban form: Evidence from five British cities. *Environment and Planning A* 41(9): 2125–2142.

Bramley G, Brown C, Dempsey N, *et al.* (2010) Social acceptability. In: Jenks M and Jones C (eds), *Dimensions of the Sustainable City*. Dordrecht: Springer, pp. 105–128.

Bunnell T and Kathiravelu L (2016) Extending urban liveability: Friendship and sociality in the lives of low-wage migrants. *International Development Planning Review* 38(2): 201–220.

Burchardt T, Grand JL, and Piachaud D (2002a) Degrees of exclusion: Developing a dynamic, multidimensional measure. In: Hills J, Grand JL, and Piachaud D (eds), *Understanding Social Exclusion*. Oxford, New York: Oxford University Press, pp. 30–43.

Burchardt T, Grand JL, and Piachaud D (2002b) Introduction. In: Hills J, Grand JL, and Piachaud D (eds), *Understanding Social Exclusion*. Oxford, New York: Oxford University Press, pp. 1–12.

Burford G, Hoover E, Velasco I, *et al.* (2013) Bringing the 'missing pillar' into sustainable development goals: Towards intersubjective values-based indicators. *Sustainability* 5(7): 3035–3059.

Burgess R (2000) The compact city debate: A global perspective. In: Jenks M and Burgess R (eds), *Compact Cities, Sustainable Urban Forms for Developing Countries*. London, New York: Spon Press, pp. 9–24.

Burton E (2000) The compact city: Just or just compact? A preliminary analysis. *Urban Studies* 37(11): 1969–2006.

Campbell H (2012) Planning to change the world: Between knowledge and action lies synthesis. *Journal of Planning Education and Research* 32(2): 135–146.

Campbell S (1996) Green cities, growing cities, just cities?: Urban planning and the contradictions of sustainable development. *Journal of the American Planning Association* 62(3): 296–312.

Chan E and Lee GKL (2007) Critical factors for improving social sustainability of urban renewal projects. *Social Indicators Research* 85(2): 243–256.

Chiu RLH (2002) Social equity in housing in the Hong Kong Special Administrative Region: A social sustainability perspective. *Sustainable Development* 10(3): 155–162.

Chiu RLH (2003a) Social sustainability, sustainable development and housing development. In: Forrest R and James Lee (eds), *Housing and Social Change: East–West Perspectives*. London and New York: Routledge, pp. 221–239.

Cilliers EJ (2018) The undervaluation, but extreme importance, of social sustainability in South Africa. In: *Smart Futures, Challenges of Urbanisation, and Social Sustainability*. Cham: Springer, pp. 131–147.

Colantonio A (2009) Social sustainability: A review and critique of traditional versus emerging themes and assessment methods. In: Malcolm H, Price A, Bebbington J, *et al.* (eds), *Proceedings of the Second International Conference on Whole Life Urban Sustainability and its Assessment.* Loughborough: Loughborough University, pp. 865–885.

Colantonio A and Dixon T (2011) *Urban Regeneration and Social Sustainability: Best Practice from European Cities.* Chichester: Wiley-Blackwell.

Coleman JS (1988) Social capital in the creation of human capital. *American Journal of Sociology* 94: 95–120.

Connelly S (2007) Mapping sustainable development as a contested concept. *Local Environment* 12(3): 259–278.

Cuthill M (2010) Strengthening the 'social' in sustainable development: Developing a conceptual framework for social sustainability in a rapid urban growth region in Australia. *Sustainable Development* 18(6): 362–373.

Dale A and Newman L (2010) Social capital: A necessary and sufficient condition for sustainable community development? *Community Development Journal* 45(1): 5–21.

Dave S (2010) High urban densities in developing countries: A sustainable solution? *Built Environment* 36(1): 9–27.

Dave S (2011) Neighbourhood density and social sustainability in cities of developing countries. *Sustainable Development* 19(3): 189–205.

Davidson M (2009) Social sustainability: A potential for politics? *Local Environment* 14(7): 607–619.

Davidson M (2010) Social sustainability and the city. *Geography Compass* 4(7): 872–880.

Davoudi S and Atkinson R (1999) Social exclusion and the British planning system. *Planning Practice and Research* 14(2): 225–236.

de Haan FJ, Ferguson BC, Adamowicz RC, *et al.* (2014) The needs of society: A new understanding of transitions, sustainability and liveability. *Technological Forecasting and Social Change* 85: 121–132.

Dempsey N, Brown C, and Bramley G (2012) The key to sustainable urban development in UK cities? The influence of density on social sustainability. *Progress in Planning* 77(3): 89–141.

Dennis RE, Williams W, Giangreco MF, *et al.* (1993) Quality of life as context for planning and evaluation of services for people with disabilities. *Exceptional Children* 59(6): 499–512.

Department of Infrastructure and Transport (2011) *Our Cities, Our Future. A National Urban Policy for Productive, Sustainable, and Liveable Future.* Canberra, Australia: Department of Infrastructure and Transport.

Dillard J, Dujon V, and King MC (eds) (2009) *Understanding the Social Dimension of Sustainability.* New York: Routledge.

Dissart J-C and Deller SC (2000) Quality of life in the planning literature. *Journal of Planning Literature* 15(1): 135–161.

Dujon V (2009) In the absence of affluence. The struggle for social sustainability in the third world. In: Dillard J, Dujon V, and King MC (eds), *Understanding the Social Dimension of Sustainability.* New York: Routledge, pp. 122–136.

Duxbury NA and Jeannotte MS (2010) From the bottom-up: Culture in community sustainability planning. In: *ESA Research Network Sociology of Culture Midterm Conference: Culture and the Making of Worlds*, Milan, 2010. Available at: http://papers.ssrn.com/abstract=1692105 (accessed 19 September 2016).

Eizenberg E and Jabareen Y (2017) Social sustainability: A new conceptual framework. *Sustainability* 9(1): 68.

Elkin T, McLaren D, and Hillman M (1991) *Reviving the City: Towards Sustainable Urban Development.* London: Continuum International Publishing.

Faber DR and McCarthy D (2003) Neo-liberalism, globalization and the struggle for ecological democracy. In: Agyeman J, Bullard RD, and Evans B (eds), *Just Sustainabilities: Development in an Unequal World.* Cambridge, MA: The MIT Press, pp. 38–63.

Fleischer M, Fuhrmann M, Haferburg C, *et al.* (2013) 'Festivalisation' of urban governance in South African cities: Framing the urban social sustainability of mega-event driven development from below. *Sustainability* 5(12): 5225–5248.

Forrest R and Kearns A (1999) *Joined-up Places: Social Cohesion and Neighbourhood Regeneration.* York: Joseph Rowntree Foundation.

Forrest R and Kearns A (2001) Social cohesion, social capital and the neighbourhood. *Urban Studies* 38(12): 2125–2143.

Friedkin NE (2004) Social cohesion. *Annual Review of Sociology* 30(1): 409–425.

Godschalk DR (2004) Land use planning challenges: Coping with conflicts in visions of sustainable development and livable communities. *Journal of the American Planning Association* 70(1): 5–13.

Goldsmith E and Allen R (1972) *A Blueprint for Survival.* Boston, MA: Houghton Mifflin.

Gough J and Eisenschitz A (2006) *Spaces of Social Exclusion.* New York: Routledge.

Grieco M (2015) Social sustainability and urban mobility: Shifting to a socially responsible pro-poor perspective. *Social Responsibility Journal* 11(1): 82–97.

Grootaert C and Bastelaer T van (2001) *Understanding and Measuring Social Capital: A Synthesis of Findings and Recommendations from the Social Capital Initiative.* Social Capital Initiative Working Paper No. 24. Washington, DC: The World Bank.

Hamiduddin I (2015) Social sustainability, residential design and demographic balance: Neighbourhood planning strategies in Freiburg, Germany. *Town Planning Review* 86(1): 29–52.

Harvey D (2009) *Social Justice and the City.* Athens, London: University of Georgia Press.

Hawkes J (2001) *The Fourth Pillar of Sustainability. Cultures Essential Role in Public Planning.* Melbourne: Common Ground Publishing Pty Ltd.

Healey P (2012) The universal and the contingent: Some reflections on the transnational flow of planning ideas and practices. *Planning Theory* 11(2): 188–207.

Kaal H (2011) A conceptual history of livability: Dutch scientists, politicians, policy makers and citizens and the quest for a livable city. *City* 15(5): 532–547.

Karen O, Lambour G, and Greenspan S (1990) Persons in transition. In: Schalock RL and Bogale MJ (eds), *Quality of Life: Perspectives and Issues.* Washington, DC: American Association of Mental Retardation, pp. 85–92.

Karuppannan S and Sivam A (2011) Social sustainability and neighbourhood design: An investigation of residents' satisfaction in Delhi. *Local Environment* 16(9): 849–870.

Kay A (2005) Territorial justice and devolution. *British Journal of Politics & International Relations* 7(4): 544–560.

Koning J (2002) Social sustainability in a globalizing world: Context, theory and methodology explored. In: van Rinsum H, de Ruijter A, Kazancigil YK, *et al.* (eds), *More on MOST: Proceedings of an Expert Meeting.* Amsterdam: UNESCO Centre, pp. 63–89.

Križnik B (2018) Transformation of deprived urban areas and social sustainability: A comparative study of urban regeneration and urban redevelopment in Barcelona and Seoul. *Urbani izziv* 29(1): 83–95.

Kytta M, Broberg A, Haybatollahi M, *et al.* (2016) Urban happiness: Context-sensitive study of the social sustainability of urban settings. *Environment and Planning B: Planning and Design* 43(1): 34–57.

Lamendola WF and Pellegrini RV (1979) Quality of life and coronary artery bypass surgery patients. *Social Science & Medicine. Part A: Medical Psychology & Medical Sociology* 13: 457–461.

Landorf C (2011) Evaluating social sustainability in historic urban environments. *International Journal of Heritage Studies* 17(5): 463–477.

Leal Filho W, Platje J, Gerstlberger W, *et al.* (2016) The role of governance in realising the transition towards sustainable societies. *Journal of Cleaner Production* 113: 755–766.

Littig B and Griessler E (2005) Social sustainability: A catchword between political pragmatism and social theory. *International Journal of Sustainable Development* 8(1/2): 65–79.

Madanipour A (1998) Social exclusion and space. In: Madanipour A, Cars G, and Allen J (eds), *Social Exclusion in European Cities: Processes, Experiences and Responses.* London: Routledge, pp. 75–89.

Magis K and Shinn C (2009) Emergent principles of social sustainability. In: Dillard J, Dujon V, and King MC (eds), *Understanding the Social Dimension of Sustainability.* New York: Routledge, pp. 15–44.

Mandzuk LL and McMillan DE (2005) A concept analysis of quality of life. *Journal of Orthopaedic Nursing* 9(1): 12–18.

Manzi T, Lucas K, Jones TL, *et al.* (eds) (2010) *Social Sustainability in Urban Areas: Communities, Connectivity and the Urban Fabric.* London: Routledge.

Massam BH (2002) Quality of life: Public planning and private living. *Progress in Planning* 58(3): 141–227.

McKenzie S (2004) *Social Sustainability: Towards Some Definitions.* Hawke Research Institute Working Paper Series 27. Magill: Hawke Research Institute.

Meadows D, Meadows DL, Randers J, *et al.* (1972) *The Limits to Growth.* New York: Universe Books.

Moons P, Budts W, and De Geest S (2006) Critique on the conceptualisation of quality of life: A review and evaluation of different conceptual approaches. *International Journal of Nursing Studies* 43(7): 891–901.

Murphy K (2012) The social pillar of sustainable development: A literature review and framework for policy analysis. *Sustainability: Science, Practice, & Policy* 8(1): 15–29.

Næss P (1994) Normative planning theory and sustainable development. *Scandinavian Housing and Planning Research* 11(3): 145–167.

Namazi-Rad M-R, Perez P, Berryman M, *et al.* (2016) A semi-empirical determination of perceived liveability. *Bulletin of Sociological Methodology/Bulletin de Méthodologie Sociologique* 129(1): 5–24.

Neuman M (1998) Does planning need the plan? *Journal of the American Planning Association* 64(2): 208–220.

Newton PW (2012) Liveable and sustainable? Socio-technical challenges for twenty-first-century cities. *Journal of Urban Technology* 19(1): 81–102.

Opp SM (2016) The forgotten pillar: A definition for the measurement of social sustainability in American cities. *Local Environment* 22(3): 286–305.

Opp SM and Saunders KL (2013) Pillar talk: Local sustainability initiatives and policies in the united states–finding evidence of the 'three e's': Economic development, environmental protection, and social equity. *Urban Affairs Review* 49(5): 678–717.

Peterson N (2016) Introducing to the special issues on social sustainability: Integration, context, and governance. *Sustainability: Science, Practice, & Policy* 12(1).

Pierson J (2010) *Tackling Social Exclusion*. Social work skills series. London: Routledge.

Polese M and Stren R (eds) (2000) *The Social Sustainability of Cities: Diversity and the Management of Change*. Toronto: University of Toronto Press.

Psarikidou K and Szerszynski B (2012) Growing the social: Alternative agrofood networks and social sustainability in the urban ethical foodscape. *Sustainability: Science, Practice & Policy* 8(1): 30–39.

Putnam RD (2000) *Bowling Alone: The Collapse and Revival of American Community*. New York, London: Simon & Schuster.

Putnam RD, Leonardi R, and Nanetti RY (1993) *Making Democracy Work: Civic Traditions in Modern Italy*. Princeton, NJ: Princeton University Press.

Raman S (2010) Designing a liveable compact city. Physical forms of city and social life in urban neighbourhoods. *Built Environment* 36(1): 63–80.

Ročak M, Hospers G-J, and Reverda N (2016) Searching for social sustainability: The case of the shrinking city of Heerlen, the Netherlands. *Sustainability* 8(4): 382.

Rogge N, Theesfeld I, and Strassner C (2018) Social sustainability through social interaction: A national survey on community gardens in Germany. *Sustainability* 10(4): 1085.

Romney DM, Brown RI, and Fry PS (1994) Improving the quality of life: Prescriptions for change. *Social Indicators Research* 33(1–3): 237–272.

Sachs I (1999) Social sustainability and whole development: Exploring the dimensions of sustainable development. In: Becker E and Jahn T (eds), *Sustainability and the Social Sciences*. London, New York: Zed Books, pp. 25–36.

Schuessler KF and Fisher GA (1985) Quality of life research and sociology. *Annual Review of Sociology* 11(1): 129–149.

Seghezzo L (2009) The five dimensions of sustainability. *Environmental Politics* 18(4): 539–556.

Serag El Din H, Shalaby A, Farouh HE, *et al.* (2013) Principles of urban quality of life for a neighborhood. *HBRC Journal* 9(1): 86–92.

Shirazi MR and Keivani R (2017) Critical reflections on the theory and practice of social sustainability in the built environment: A meta-analysis. *Local Environment* 22(12): 1526–1545.

Shirazi MR and Keivani R (2018) The triad of social sustainability: Defining and measuring social sustainability of urban neighbourhoods. *Urban Research & Practice* 0(0): 1–24.

Smith A (2009) Theorising the relationship between major sport events and social sustainability. *Journal of Sport & Tourism* 14(2–3): 109–120.

Smith A (2010) The relationship between major events, the urban fabric and social sustainability. In: Manzi T, Lucas K, Lloyd-Jones T, *et al.* (eds), *Social Sustainability in Urban Areas. Communities, Connectivity and the Urban Fabric*. London: Earthscan, pp. 199–219.

Social Exclusion Unit (1998) *Bringing Britain Together: A National Strategy for Neighbourhood Renewal*. London: HMSO.

Søholt S, Ruud ME, and Braathen E (2012) A question of social sustainability: Urban interventions in critical neighbourhoods in Portugal and Norway. *Urban Research & Practice* 5(2): 256–272.
Soini K and Birkeland I (2014) Exploring the scientific discourse on cultural sustainability. *Geoforum* 51: 213–223.
Soleimani M, Tavallaei S, Mansuorian H, *et al.* (2014) The assessment of quality of life in transitional neighborhoods. *Social Indicators Research* 119(3): 1589–1602.
Southworth M (2003) Measuring the liveable city. *Built Environment* 29(4): 343–354.
Stein SM and Harper TL (2005) Rawls's 'justice as fairness': A moral basis for contemporary planning theory. *Planning Theory* 4(2): 147–172.
Stren R and Polese M (2000) Understanding the new sociocultural dynamics of cities: Comparative urban policy in a global context. In: Polese M and Stren R (eds), *The Social Sustainability of Cities: Diversity and the Management of Change*. Toronto: University of Toronto Press, pp. 13–38.
Szalai A (1980) The meaning of comparative research on the quality of life. In: Szalai A and Andrews FM (eds), *The Quality of Life: Comparative Studies*. London: Sage Publications.
Taket A, Crisp BR, Nevill A, *et al.* (2009) *Theorising Social Exclusion*. London, New York: Routledge.
Turner J (2012) Urban mass transit, gender planning protocols and social sustainability: The case of Jakarta. *Research in Transportation Economics* 34(1): 48–53.
Turok I, Kearns A, and Goodlad R (1999) Social exclusion: In what sense a planning problem? *The Town Planning Review* 70(3): 363–384.
Vallance S, Perkins HC, and Dixon JE (2011) What is social sustainability? A clarification of concepts. *Geoforum* 42(3). Themed Issue: Subaltern Geopolitics: 342–348.
Vuchic V (1999) *Transportation for Livable Cities*. New Brunswick, NJ: Routledge.
Weingaertner C and Moberg Å (2014) Exploring social sustainability: Learning from perspectives on urban development and companies and products. *Sustainable Development* 22(2): 122–133.
Wheeler SM (2013) *Planning for Sustainability: Creating Livable, Equitable and Ecological Communities*. London, New York: Routledge.
Whitehead M (2011) The sustainable city: An obituary? On the future for and prospects of sustainable urbanism. In: Flint J and Raco M (eds), *The Future of Sustainable Cities: Critical Reflections*. Bristol: Policy Press, pp. 29–46.
Winkler T and Duminy J (2016) Planning to change the world? Questioning the normative ethics of planning theories. *Planning Theory* 15(2): 111–129.
Winston N and Pareja Eastaway M (2008) Sustainable housing in the urban context: International sustainable development indicator sets and housing. *Social Indicators Research* 87(2): 211–221.
Woodcraft S (2012) Social sustainability and new communities: Moving from concept to practice in the UK. *Social and Behavioral Sciences* 68: 30–42.
World Bank (1998) *The Initiative on Defining, Monitoring and Measuring Social Capital*. Social Capital Initiative Working Paper 1. World Bank. Available at: http://siteresources.worldbank.org/INTSOCIALCAPITAL/Resources/Social-Capital-Initiative-Working-Paper-Series/SCI-WPS-01.pdf [accessed 20 August 2018].
World Commission on Environment and Development (1987) *Our Common Future*. Oxford: Oxford University Press.

World Health Organization (1995) WHOQOF-100. Geneva: WHO. Available at: www.who.int/healthinfo/survey/WHOQOL-100.pdf?ua=1 (accessed 30 March 2017).

Yiftachel O and Hedgcock D (1993) Urban social sustainability: The planning of an Australian city. *Cities* 10(2): 139–157.

Yoo C and Lee S (2016) Neighborhood built environments affecting social capital and social sustainability in Seoul, Korea. *Sustainability* 8(12): 1346.

Yung HKW and Chan HWE (2012) Critical social sustainability factors in urban conservation: The case of the central police station compound in Hong Kong. *Facilities* 30(9/10): 396–416.

Åhman H (2013) Social sustainability – Society at the intersection of development and maintenance. *Local Environment* 18(10): 1153–1166.

2 Social sustainability

Politics and democracy in a time of crisis

Mark Davidson

Introduction

Does social sustainability carry any political meaning? The answer to this question is not obvious. Since the 1990s, sustainability has proven a rallying call for policy makers and activists of many stripes. Emerging alongside a growing scientific concern with climate change, sustainability has transformed contemporary political and policy thought. As such, it is unlikely to go away any time soon. As the triple-bottom-line (TBL) conceptualization of sustainability (Elkington, 1999) has gained widespread acceptance, public debates have become more focused on developing programmes for sustainability's constitutive elements. If we evaluate the effectiveness of the concept of sustainability in terms of popularity – including TBL variants – it appears to have been a highly effective idea that is commonly used to think about various forms of change.

Yet, we can claim that sustainability has utterly failed. Although it has changed the way people think about environmental, social and economic issues, it is often charged with only marginally influencing material processes (e.g. Swyngedouw, 2010). If we measure the success of sustainability agendas against the initial premise – that we need to create a more sustainable environmental relationship between Planet Earth and humans (Bruntland, 1987) – perhaps all we can claim to have done is heighten our awareness about unprecedented climatic trauma (Crutzen, 2006). Of course, sustainability is only a concept and it is unrealistic to expect that an idea alone can transform complex human–environment relationships. We may simply need more time for action to be stimulated by new ways of thinking.

Others see the situation differently. Some now consider sustainability an archetypal element of what is referred to as the 'post-political' (Swyngedouw, 2010). The primary charge made against sustainability is that it has failed to mobilize necessary political change and perpetuated the status quo. It has generated a situation whereby everyone is for sustainability, but few have mobilized change (ibid.). This combination of widespread agreement and absence of social change is the opposite of 'politics' for Jacques Rancière (1999). Politics, in the Rancière (1999) formulation, is premised on the production of disagreement. Politics involve the pitting of one world against another, and is are about a certain type of demand that transforms the world from which the demand emanates. It is here that sustainability is open to critique. One has only to compare language used by Greenpeace[1] and British Petroleum[2] to see the problem: both are strong advocates of sustainability, thus showing that the concept does not seem to realize the disagreements that clearly exist between the two parties.

Without clear political content (i.e. disagreement), the utility of the (social) sustainability concept might well be in doubt. In theoretical terms, we might be witnessing the end of sustainability as an organizing concept. This is not to say that the word sustainability will simply disappear. The immediate associational discursive power of sustainability, i.e. its ideological operation (Žižek, 1989; Laclau, 2006), will remain important. Its appeal to notions of environmental stability, intergenerational equity, long-term solutions and holistic thinking will ensure that it remains an important political and policy idea. Rather, I am arguing that these types of associational meanings are coming under increasing pressure, to the extent that sustainability may be a concept now used in largely cynical ways.

In this chapter, I therefore want to define and examine the political meaning of social sustainability. I begin by examining some of the different iterations of social sustainability that have been developed across the academic literature. This is not intended as a comprehensive review of the existing uses of social sustainability. Rather the goal is to highlight some of the political implications of existing definitions and develop a critical interpretation of usage variance itself. By identifying the multiple and varied usage of 'social sustainability', the chapter sets up the problem of how a seemingly fuzzy concept comes to operate in a highly popular and influential manner. I answer this by arguing that social sustainability is an empty signifier (Davidson, 2010a). From this perspective, we can then understand why (social) sustainability seems so widely accepted but often lacks accompanying transformative action. Sustainability might therefore be an exemplary concept of what Slavoj Žižek (2011) calls 'the end times'. However, I want to argue that this need not be the case. Sustainability and, in particular, social sustainability, needs to become more closely linked with the idea of democracy. Whether in the mould of Rancière's (1999) radical tradition or Dewey's (1989) pragmatism, social sustainability needs to become political in direct reference to democratic equality.

Defining the political meaning of social sustainability

A broad societal embrace of social sustainability stems from the recognition that environmental sustainability cannot be divorced from broader processes (Bruntland, 1987). Most famously, the 1987 Brundtland Commission report rejected the idea that socio-environmental problems can be neatly separated:

> Until recently, the planet was a large world in which human activities and their effects were neatly compartmentalized within nations, within sectors (energy, agriculture, trade), and within broad areas of concern (environment, economics, social). These compartments have begun to dissolve. This applies in particular to the various global 'crises' that have seized public concern, particularly over the past decade. These are not separate crises: an environmental crisis, a development crisis, an energy crisis. They are all one.
>
> (Bruntland, 1987: 10)

The concept of sustainability that emerged from this period is one that insisted on interconnections and a rejection of silo thinking. The TBL conceptualizations of sustainability (Elkington, 1999) that subsequently developed have generated a heightened awareness that ecological problems cannot be solved without an insistence on economic and social sustainability. Yet, despite the obvious benefits of interconnected concepts like TBL sustainability, it does not remove the need to define just what the constituent elements of the TBL approach are. Unfortunately, or perhaps by necessity, these efforts to define the constitutive elements of TBL sustainability – environmental, economic and social (and sometimes cultural) – often result in some decoupling of the constituting elements.

There is now an extensive literature that attempts to define social sustainability (Dempsey *et al.*, 2011). In their examination of social sustainability, Polèse and Stren (2000) developed an understanding of social sustainability that can stand independently of any concern with the environmental and economic components of the TBL:

> development (and/or growth) that is compatible with the harmonious evolution of civil society, fostering an environment conducive to the compatible cohabitation of culturally and socially diverse groups while at the same time encouraging social integration, with improvements in the quality of life for all segments of the population.
>
> (Polèse and Stren, 2000: 16–17)

In this formulation, social sustainability concerns harmony, cohabitation, diversity, integration and distributional justice. The vision presented is progressive, setting up a sustainable society as one that respects difference, enables social mixing and delivers growing prosperity to all. When placed in the context of cities, Yiftachel and Hedgcock's (1993: 139–140) search for a definition comes to a similar conclusion, arguing that a socially sustainable city is one that involves interaction and development across all groups.

Polèse and Stren's (2000) definition helps to demonstrate some of the conceptual separation that occurs when the constituent elements of the TBL are individually defined. Unlike environmental sustainability, where normative or political judgements can be avoided by an appeal to global climate degradation (Bruntland, 1987; Swyngedouw, 2010), social sustainability seems to move quickly into normative territory. Furthermore, social sustainability lacks the apparent meaning of environmental sustainability. Public understandings of environmental sustainability are often connected to climate change science (Crutzen, 2006; Brulle *et al.*, 2012). As such, the concept of environmental sustainability carries a great deal of implicit content. For example, it is easily assumed that environmental sustainability policies work towards making our environmental conditions compatible with the foreseeable inhabitation of Earth by humans. The political content of the concept is muted by appeals to issues such as extinction and intergenerational equity. The same implicit content often cannot be identified with social sustainability (see Marcuse, 1998; Maloutas, 2003).

Definitions of social sustainability therefore tend to be weaker compared with the TBL counterparts. For example, from his excellent review of the social sustainability literature, McKenzie (2004: 12) developed the following conceptualization: 'a life-enhancing condition within communities, and a process within communities that can achieve that condition'. Although straightforward enough, the definition begs the question of what is 'life enhancing', and what might a life-enhancing condition for one community mean for another? Any number of justice theories might be applied to this problem, from Aristotle's 'good life' through to John Rawls' (1999) disinterested liberalism. Maloutas (2003) is therefore correct to argue that social sustainability tends to lack any consistent normative message. Consequently, Maloutas (2003) has argued that social sustainability tends to be subservient to other TBL components. Environmental sustainability initiatives, he argues, are often stripped down to social programming goals in order that the environmental reforms can be implemented. Political calculation can therefore trump any commitment to TBL sustainability. Due to this political expediency, Maloutas (2003: 168) goes on to argue that the normative content of social sustainability usually takes a conservative form. Although some theorists and politicians might take the idea of 'life enhancing' to suggest radical social reform, what usually happens is that appeals to less prescriptive ideas such as inclusion and cohesion are preferred.

Definitions of social sustainability might therefore be restrained by the environmental component of TBL sustainability. Yiftachel and Hedgcock (1993) took just this approach. They argued that 'the concept of urban social sustainability conceives the city as a backdrop for lasting and meaningful social relations that meet the social needs of present and future generations' (ibid.: 140). Borrowing from environmental debates and the associated concern of intergenerational equity, this formulation looks at social sustainability as a social quality. Indeed, one can imagine the construction of certain empirical tests that project out social trends to forecast some kind of unsustainable breaking point. Think, for example, of Thomas Piketty's (2013) headline-grabbing work on the relationship between capitalist wealth creation and social inequality. Using Yiftachel and Hedgcock's (1993) definition, we might transform Piketty's now famous graphs of widening inequality into a social sustainability concern: continuing inequality trends do not permit the meeting of future social needs, hence the situation is socially unsustainable.

Here we can begin to see how the recent concern with social sustainability might relate to more longstanding political debates. Bahler (2007) argues that social sustainability is usually a concept that just reworks longstanding social problems. These are the types of problems that democracy emerged as a solution so: 'we might venture to define social and political (or 'nationhood') sustainability as the ability of a society to resist internal forces of decay while also maintaining and reproducing the background social, cultural, and institutional conditions necessary for healthy democratic social relations to flourish' (Bahler, 2007: 27). Whether or not a democratic social arrangement and associated institutions would survive was, of course, a central concern of nineteenth-century political commentators like de Tocqueville (see Bahler, 2007). Although the term 'social sustainability' was not used by nineteenth-century political commentators, it is clear that de Tocqueville's commentary on the early American democratic experiment is motivated, in large part, by the question of whether or not it would persist. De Tocqueville saw early American appeals of equality and democratic government to be powerful enough to persist; however, he became concerned with whether the situation in emergence, popular representative democracy, would be worth sustaining. De Tocqueville's comments predicted a sustainable but undesirable situation (De Tocqueville, 2000 [1840]):

> It does not break men's will, but softens, bends, and guides it; it seldom enjoins, but often inhibits, action; it does not destroy anything, but prevents much being born; it is not at all tyrannical, but it hinders, restrains, enervates, stifles, and stultifies so much that in the end each nation is no more than a flock of timid and hardworking animals with the government as its shepherd.

By placing our concern with social sustainability within this context, some of the contradictions of the term become more apparent. Most theorists of social sustainability have argued that sustainable societies embrace equality, integration and distributional justice. All fair enough, but what underpins our commitment to these values, and why? Reading de Tocqueville into this discussion makes us aware that the idea of social sustainability brings us very close to the original questions of political philosophy: What is the good life? What is a good society? And as de Tocqueville's own investigations of Jefferson's emerging democratic experiment demonstrated, just because a society might be attempting to achieve noble goals does not mean it is, by definition, sustainable. We can therefore find ourselves in a circular inquiry: When we aim to make our society sustainable, we make it a requirement to identify what type of society we want. Yet when we do this, there is no guarantee that we come closer to knowing if it is sustainable or worth sustaining. It is therefore necessary to think more about what kind of conceptual work we need social sustainability to perform.

Conceptualizing social sustainability as an empty signifier

The academic literature's uncertainty (Dempsey *et al.*, 2011) about the concept of social sustainability is not mirrored in the policy world. Here, the idea of social sustainability is widespread, and many policy programmes are formulated around the concept (Davidson, 2010b). This appears to be a strange situation. Perhaps we are faced with multiple versions of social sustainability that will, eventually, become a coherent concept. Or perhaps social sustainability is yet another 'fuzzy concept' that varies dramatically depending on the context to which it is applied. In this section I suggest that social sustainability – as part of sustainability more generally – operates as an empty signifier (Laclau, 2006; Davidson, 2010a). An empty signifier is a term that performs an organizing duty within a social discourse (i.e. ideology) but lacks any definitive content itself. For Barthes ([1957] 2011), the empty signifier is without any certain signified. It is purely a pivotal point, something used to orientate a set of other concepts (Gunder, 2004).

Where social problems that might previously have been framed by 'social injustice', 'inequality' or 'deviance', for example, they now find themselves organized around the idea of sustainability. As an empty signifier, social sustainability can function to turn disordered thoughts into coherent understanding (see Žižek, 1989 [2009]). There are many conceptual implications involved in understanding (social) sustainability in this way (see Davidson, 2010a). Here I want to highlight just one, relating to what political philosopher Ernesto Laclau (2006) called the nominal status of empty signifiers.

Empty signifiers such as social sustainability are distinguished from other concepts by the fact they have a nominal status:

> It is a name.... If the various determining components of an object shared some essential features preceding the act of naming it, the act of naming would be ancillary to a conceptual mediation. But if those features are heterogeneous and, as a result, radically contingent, the unity of the object has no other ground than the act of naming it.
>
> (Laclau, 2006: 109)

Laclau (2006) is here arguing that many of the ideas with which we organize our collective understandings are nothing but names. These names are distinguished from concepts because they bring with them certain organizing principles. If (social) sustainability were therefore a fuzzy concept (Markusen, 2003), you would be able to see certain types of understanding within various manifestations. This creates the opportunity for refinement. For example, you could take multiple understandings and/or applications of social sustainability and seek to produce a less fuzzy concept over time (ibid.). In contrast, an empty signifier is much more unstable and can therefore carry with it radically different meanings. Although this nominal, empty status might first appear to be a damning weakness, philosophers such as Laclau (2006) and Žižek (1989) have attempted to demonstrate how empty signifiers perform foundational ideological functions.

The nominal purpose of the empty signifier is reflective, performing a kind of stoppage in the prevailing ideology. One can think of this in terms of the multiple possible meanings of most concepts. If all meanings were constantly in flux, the ideology would not perform any useful social function. Empty signifiers therefore act to quilt the free-floating ideological elements, making them into a 'structured network of meaning' (Žižek, 1989: 87). Coherence is therefore delivered by naming, not a sublimation of meaning by an ordering concept. In the former, a mastering concept would deliver meaning to its secondary concepts. A nominal, empty signifier functions in the opposite manner, having no immediate conceptual export. Rather, meaning and understanding are generated by the way the act of naming itself organizes secondary concepts into relations, enabling meaningful content to flow back to the nominal.

Throughout his philosophical works, Žižek (1989) has attempted to demonstrate the implications of this understanding of empty signifiers and ideology. A core part of this effort has been informed by the psychoanalytical theory of Jacques Lacan. Žižek argues that empty signifiers are voids, names with no inherent content. In respect to this discussion, we can say social policies are often organized around the idea of social sustainability, but that the idea of social sustainability itself has little positive content. One unavoidable issue with this arrangement is that the void of the nominal takes on a traumatic quality. Borrowing from psychoanalysis, Žižek argues that the importance of the naming process is always clear: it brings coherence. However, if we probe too much into what the name itself contains, then a traumatic experience can occur. This can be illustrated with a couple of examples.

First, imagine a psychotherapy group (see Urban, 2008). The whole purpose of this type of therapy is that a group setting enables certain kinds of healing to occur. Everyone therefore shows up at the therapy session assuming a 'group' is there, and that they will interact with the group in a way that is therapeutic. However, the group does not exist in any positive sense. Rather, a group of individuals comes together, all assuming that something with the qualities of a group exists in order to access the therapy. The group exists only in so much as the individual members assume it exists. If members of the group started to try to understand what the 'group' is, they would probably find themselves with different understandings, potentially demonstrating that there is nothing that can be substantially described as 'the group'. Hence, a disorienting trauma may ensue, and the therapy breaks down. Yet, key to this example is that, despite 'the group' being nothing in a positive sense, it still functions as a constructive therapeutic device.

Transfer this same understanding to the second example. Here we are in a policy-making process where a group of people are attempting to deliver more socially sustainable outcomes. Everyone is agreed on the idea, or at least is being paid to develop socially sustainable policies, and they start by examining various areas of policy: housing, unemployment benefit, vocational training, and so on. They adopt a particular understanding of social sustainability based on something like McKenzie's (2004) life-enhancing condition. But then someone starts to question what the idea of social sustainability actually means? Is it about sustaining the existing society, or building a sustainable society? What is sustainable anyway, and does sustainability mean unnecessarily preserving certain social processes? Given that social sustainability is itself empty of definitive content, and able to be defined only by a set of unstable relational concepts, we end up in a similarly traumatic process to the first example. The web of meaning we have used to orientate our social and individual actions can begin to dissolve, generating disorienting trauma.

Žižek (2004) describes this as the empty signifier always ultimately failing. The concepts that we collectively use to organize our social actions are therefore always on the edge of collapse. Drawing from post-structuralist linguistic theory (Torfing, 1999), Žižek's theorization is important for understanding social sustainability because it recognizes the necessity of empty signifiers in our ideological makeup, but also describes their untameable constitution. They are both indispensable and inherently problematic. We should therefore recognize that social sustainability definitions are multiple and that they all rely on the induction of secondary concepts. We should also recognize that any attempt to produce a master conceptualization of social sustainability is probably doomed from the start. This would be, for Žižek, the dream of a post-ideological time. So, where do we go from here?

Social sustainability in a time of crisis

Understanding social sustainability as an empty term does not mean that it is useless. Most of the terms we use to coordinate social action – equality, justice, community – have a similar constitution (Laclau, 2006). What we must do is critically examine how useful a particular construct, such as social sustainability, is at any one time. This position is familiar for any reader of the pragmatist philosophy of John Dewey (2008 [1925]). Dewey (2008 [1925]), although coming from a different philosophical position to Laclau or Žižek, argued against the idea that we could secure transcendental concepts to anchor human action. Setting up his post-foundational position, Dewey argued that we must always revisit how our intellectual frameworks measure up as interpretative devices:

> It warns us that all intellectual terms are the products of discrimination and classification, and that we must, as philosophers, go back to the primitive situations of life that antecede and generate these reflective

> interpretations, so that we re-live former processes of interpretation in a wary manner, with eyes constantly upon the things to which they refer. Thus empiricism is the truly critical method; it puts us knowingly and cautiously through steps which were first taken uncritically, and exposed to all kinds of adventitious influence.
>
> (Dewey, 2008 [1925], p. 386)

Dewey (2008 [1925]) is arguing for a critical, reflective and empirically informed method of conceptual development. He argued we must always be careful not to use our conceptual schemes in a dogmatic or unreflective manner, otherwise we risk using abstractions that distort and degrade our engagements with the world. To somewhat over-simplify, Dewey wanted philosophy to continually ask how useful it was to the achievement of human ends. If we return to social sustainability, a recognition of its empty signifier status need not trap us in some nihilistic or relativist position. Instead, we can interrogate how useful the concept social sustainability is at this moment in time.

We can start this task by folding the current discussion of social sustainability back into a broader concern with sustainability. For philosophers like Žižek (2011), the idea of systemic crisis is linked to a host of deep, interlocked contradictions, specifically environmental change, biogenetics, intellectual property and new social apartheids. These contradictions stand out for Žižek (2011) because their resolution remains outside of the current coordinates of liberal democratic capitalism. Nancy Fraser (2015) has offered a similar critique of our current social condition, arguing that financialised capitalism is undergoing an unprecedented legitimization crisis. Fraser (2015) is here borrowing from Jurgen Habermas's (1975) theory of legitimation crisis. Habermas (1975) argued that the intermittent crises of capitalist economies tend to be shifted into the political sphere. Rather than resolving a crisis of economic accumulation, the problem is therefore moved around, causing a host of state restructuring problems. Fraser (2015) has argued that we now face a situation where crises are not being displaced. Rather we are undergoing a set of concurrent crises in the economic, political and social spheres. A route out of the economic crisis is therefore much more difficult, and so on. Such a situation demands types of thinking that can produce radical and remedial social change.

Fraser's (2015) thesis gives us pause to assess what role the concepts of sustainability (in general) and social sustainability (specifically) play today. When sustainability emerged in the 1980s, it was an empty signifier commonly related to notions of climate stability, safe limits and homeostasis (Bruntland, 1987; Elkington, 1999). Sustainability literature at the time was principally concerned about climate change, with many accounts thinking that climatic balance might be restored. Today's literature reads differently (Crutzen, 2006). Now we are faced with questions of climate uncertainly, instability, mass extinctions, long-term refugee crises, the Anthropocene and planetary exodus. Unwittingly or not, expectations have shifted about our climatic future. As time goes by, the future we are willing to live with and pass on to our kids has changed. Descriptions of the Anthropocene symbolize a new type of thinking about human–environmental relationships that should make us reflect on whether sustainability concepts remain relevant today.

As we enter the Anthropocene (Crutzen, 2006), we are certainly moving into uncharted territories. Due to an oligarchic global political structure and the failure of our democratic institutions, we now require a radical embrace of the new climatic period (Žižek, 2011). Within the climate science community, there appears broad agreement that there is no going back. We now have two options, neither of which relates to the ideas of return, balance or predictable futures, all central associations of sustainability. Our two choices seem to be a passive move into an unpredictable climatic condition – something that moves us beyond any appeal to resilience – or a radical re-imagination of human–environmental relationships whereby things such as geo-engineered remedial solutions are used to manage the planetary system as a human technology (Wigley, 2006). The latter constitutes nothing less than a rethinking of human civilization and its relationship to Planet Earth. It also involves taking on a degree of environmental responsibility that so far has evaded the capacity of our political institutions.

Of course, we also have to relate this environmental debate to the social and economic spheres if we are to assess the total future relevance of sustainability. Here I think we find a similar story of crisis and the need for radical solutions that challenge the efficacy of sustainability thinking. In the social sphere, we see Victorian-level socioeconomic divides (Piketty, 2013) and tribalistic forms of politics (Rancière, 1999) that, by definition, deny any type of universal emancipatory dimension. This is the post-political condition where different groups compete over certain resources without any recourse to universally accepted forms of political reason (Swyngedouw, 2010). Within societies that proclaim themselves to be democratic, this universal dimension is meant to be inscribed in the equality presumption:

> Such justice only begins wherever uses stop being parceled out, wherever profits and losses stop being weighed. Justice as the basis of community has not yet come into play wherever the sole concern is with preventing individuals who live together from doing each other reciprocal wrongs and with reestablishing the balance of profits and losses whenever they do so. It only begins when what is at issue is what citizens have *in common* and when the main concern is with the way the forms of exercising and of controlling the exercising of this common capacity are divided up.
>
> (Rancière, 1999: 4–5)

A concern with what is in common is, of course, difficult to identify today. As Peter Sloterdijk (2011) has extensively described, our societies are characterized by bubbles: lives lived in separation where the idea of tolerance takes the form of a demand against harassment. 'Let me be, and do not come too close' is the radical injunction of multicultural tolerance, according to Žižek (2004).

The post-political moment is therefore paradoxical. Our societies are legitimated by an appeal to democratic equality, and yet this is manifest in a fractured and divided sociopolitical landscape where the normative assessment of social difference is resisted on a psychological and political basis. The converse of this is situation is, of course, that the emergence of the political demand that harasses – 'you must change your life' – is too easily dismissed as another round of tribalistic conflict.

Things do not look much better in the economic sphere. As Nancy Fraser (2015) argues, financialised capitalism has transmitted crises across all spheres. The consequent inability for capitalism to displace and temporarily resolve its legitimation crisis therefore appears lacking. One consequence has been the creation of a political landscape where governance appears devoid of government. States appear unable to do much to reconstruct economic processes into less unstable and more equitable forms. If Fraser (2015) is correct, the implications of capitalism's failure to displace its legitimation crisis are profound. With productivity gains hard to find in developed economies, lagging growth undermining the social model that has been erected over the past 40 years, and states appearing inept despite reforms with an overtly non-democratic form, the bargain between capitalism and liberal democracy looks to be breaking down (Brown, 2003). Questions of sustaining our economy, or even building a sustainable form of capitalism, appear to have become potentially superseded by our shared radical challenges.

This rather grim picture presents significant challenges to sustainability scholars. In part, this is due to the crisis scene that has emerged within the broad-based embrace of sustainability. It is not that sustainability thinking has been lacking, but rather that its appeals seem to have faded in comparison to more powerful processes. If we are to confront the unfolding crisis, do we therefore double-down on sustainability, pursuing the idea as a mechanism to deliver the manifold promises of the TBL approach? I fear not. As we move across the interlocking crisis of the present moment, I think we will probably require a reorganization of our ideological engagement. Sustainability will remain an important idea, but the powerful associational meaning of the idea must be supplemented with a commitment to democracy. The idea of sustainability has not enabled us to make the necessary political distinctions. To put it in Carl Schmitt's terms, sustainability has become an idea that does not demark them and us, that distinction that is the lifeblood of politics (Mouffe, 2005). My proposal is therefore that social sustainability becomes an idea that is subservient to an ancient idea: democracy (Rancière, 1999). Social sustainability must denote the ancient concern of all democrats: how do we sustain and achieve the democratic condition despite countervailing forces?

Towards a democratic social sustainability

The proposition is that making social sustainability a useful concept (Dewey, 2008 [1925]) requires linking it more explicitly with democracy. Put differently, the only legitimate form of social order to maintain/sustain is democratic (see Rancière, 1999). Here I do not mean the ballot box and elections, but rather with what Jacques Rancière (2006) describes as the logic of democracy: equality. Rancière (2006: 45) claims that: 'If politics means anything, it means something that is added to all these governments of paternity, age, wealth, force and science.' Democratic politics are founded on a 'primary limitation of the forms of authority that govern the social body' (ibid.: 45). With the absence of a title to govern, all forms of authority must be assessed with regard to political claims, i.e. disagreements over the equal status of peoples. This is why Rancière finds the notion of 'consensus democracy' a 'conjunction of contradictory terms' (Rancière, 1999: 95). The problem of social sustainability must therefore be brought into dialogue with the only legitimate mode of social ordering to become effective. It must adopt what Rancière describes as the core of politics: 'Every politics is democratic in this precise sense: not in the sense of a set of institutions, but in the sense of forms of expression that confront the logic of equality with the logic of the police order' (Rancière, 1999: 101). A commitment to democratic politics, in Rancière's (1999) sense, will have many impacts on understandings of social sustainability. I want to conclude by discussing two implications of a democratic commitment relating to the associated meaning and conceptual linkages of social sustainability.

First, a commitment to democratic theorizing will strip away connotations of balance and stasis from social sustainability and replace them with the necessity that equality is inscribed on the forever-emerging (i.e. not balanced) social form. The democratic political operation is not a demand for equality, but rather a demand premised on the equality of each with all. If governmental authority does not affirm the equality of all, then it, by definition, becomes illegitimate. Politics therefore consistently lodges one world against another: the existing order assessed against the commitment to equality. Democratic politics consists of those actions that reject existing identifications through a process of political subjectification. For Rancière (1999) democracy is therefore an ongoing process, and this form of politics is defined by the emergence and the democratic evaluation of (political) disagreements.

The power of this approach lies in its ability to insert the universal dimension of politics into particular struggles. Whereas sustainability attempts offer an all-encompassing understanding of complex, interwoven processes, a concern with democracy can be mobilized only within the particular, i.e. we can only assess the status of equality within the context of particular claims (Davidson and Iveson, 2015). As we come to live in an epoch of multiple, interlocking crises, sustainability has emerged as a response that seeks to provide an overarching approach. Rancière's (1999) formulation of democratic politics can make us question this top-down orientation. In the social sphere, change will necessarily require an assessment of disagreements. If we move to the particular, each case of disagreement, we begin to examine how a commitment to democracy changes our perspective on a host of social issues. In each particular social struggle, we can ask two questions: 'Are things sustainable?' and 'Are things democratic?'. Without an affirmative answer on the latter, the goal of making something sustainable becomes nonsensical.

The final implication of the coupling of democracy and social sustainability I want to cover here concerns the broader relationship between social sustainability and the TBL. The TBL approach to sustainability carries with it significant import from environmental issues. Given the empty signifier status of all the constituting elements, there is inevitably an influence on the content of social sustainability by the other parts of the TBL. For example, powerful environmental sustainability ideas such as ecological balance and ecosystem stability can easily transfer into ideas of neighbourhood stability and social conservatism. However, we must be cautious of unwittingly accepting this type of conceptual influence (see Dewey, 2008 [1925]). It may be that, although the popularity of social sustainability emerged from the power of the TBL approach, a search for the utility of social sustainability demonstrates the need to distinguish between the sustainability concepts. Put simply, the idea of democracy – as change and disagreement – may often be overshadowed by the demands of other TBL elements. Our commitment to the TBL theorization should therefore be subject to the same tests of any concept: is it useful to the situation we are encountering? Without this type of reflection, we have a well-meaning commitment to sustainability thinking that is ineffective in practice (Dewey, 2008 [1925]).

In this chapter, I have argued that (a) there are multiple definitions of social sustainability, (b) social sustainability is an empty signifier, (c) we must assess social sustainability based on its utility, (d) the utility of (social) sustainability is brought into question during our times of crises, and (e) we can regain its utility with a serious engagement with the idea of democracy. Each of these steps in the argument is undertaken to find a way out of the current situation: a demand for more sustainable futures in the context of limited action. With respect to social sustainability, this might involve (a) rethinking how we approach notions of social change and (b) critically interrogating the ways in which the TBL couching of social sustainability creates potentially counter-productive modes of thought. In summary, there is a great deal of empirical and conceptual needed. Yet conceptual development alone will not suffice. We must also turn our attention to those democratic institutions that are meant to enable our democratic commitments. It is in these institutions that we can find a route to a socially sustainable society in which we might want to live. As the social condition tells us, there is much work to be done.

Notes

1 See: www.greenpeace.org/usa/sustainable-agriculture/issues.
2 See: www.bp.com/en/global/corporate/sustainability.html.

References

Bahler, K. (2007). Social sustainability: New Zealand solutions for Tocqueville's problem. *Social Policy Journal of New Zealand*, 31: 22–40.

Barthes, R. (2013 [1957]). *Mythologies*. Reprint, New York: Hill & Wang.

Brown, W. (2003). Neo-liberalism and the end of liberal democracy. *Theory & Event*, 7(1).

Brulle, R. J., Carmichael, J., and Jenkins, J. C. (2012). Shifting public opinion on climate change: An empirical assessment of factors influencing concern over climate change in the U.S., 2002–2010. *Climatic Change*, 114(2): 169–188.

Bruntland, G. (ed.) (1987). *Our Common Future: The World Commission on Environment and Development*. Oxford: Oxford University Press.

Crutzen, P. J. (2006). The 'Anthropocene'. In Ehlers, E. and Krafft, T. (eds), *Earth System Science in the Anthropocene*. Berlin: Springer, pp. 13–18.

Davidson, M. (2010a). Social sustainability and the city. *Geography Compass*, 4(7): 872–880.

Davidson, M. (2010b). Sustainability as ideological praxis: The acting out of planning's master signifier. *City*, 14(4): 390–405.

Davidson, M. and Iveson, K. (2015). Recovering the politics of the city: From the 'post-political city' to a 'method of equality' for critical urban geography. *Progress in Human Geography*, 39(5): 543–559.

De Tocqueville, A. (2000 [1840]) *Democracy in America: The Complete and Unabridged Volumes I and II*. New York: Bantam Books, pp. 691–692.

Dempsey, N., Bramley, G., Power, S., and Brown, C. (2011). The social dimension of sustainable development: Defining urban social sustainability. *Sustainable Development*, 19(5): 289–300.

Dewey, J. (1997). *Democracy and Education*. New York: Free Press.

Dewey J. (2008 [1925]) *The Later Works of John Dewey, Volume 1, 1925–1953: 1925: Experience and Nature* (ed. J. A. Boydston). Carbondale, IL: Southern Illinois University Press.

Elkington, J. (1999). *Cannibals with Forks: The Triple Bottom Line of 21st Century Business*. Oxford: Capstone Publishing Ltd.

Fraser, N. (2015). Legitimation crisis? On the political contradictions of financialized capitalism. *Critical Historical Studies*, 2(2): 157–189.

Gunder, M. (2004). Shaping the planner's Ego-Ideal: A Lacanian interpretation of planning education. *Journal of Planning Education and Research*, 23(3): 299–311.

Habermas, J. (1975). *Legitimation Crisis*. Boston, MA: Beacon Press.

Laclau, E. (2006). Ideology and post-Marxism. *Journal of Political Ideologies*, 11(2): 103–114.

Maloutas, T. (2003). Promoting social sustainability: The case of Athens. *City*, 7(2): 167–181.

Marcuse, P. (1998). Sustainability is not enough. *Environment and Urbanization*, 10(2): 103–112.

Markusen, A. (2003). Fuzzy concepts, scanty evidence, policy distance: The case for rigour and policy relevance in critical regional studies. *Regional Studies*, 37(6–7): 701–717.

McKenzie, S. (2004). *Social Sustainability: Towards Some Definitions*. Hawke Research Institute. University of South Australia. Available at: http://w3.unisa.edu.au/hawkeinstitute/publications/downloads/wp27.pdf [accessed 25 August 2018].

Mouffe, C. (2005). *On the Political*. New York: Routledge.

Piketty, T. (2013 [2017]). *Capital in the Twenty-First Century*. Cambridge, MA: Belknap Press.

Polèse, M. and Stren, R. E. (2000). *The Social Sustainability of Cities: Diversity and the Management of Change*. Toronto: University of Toronto Press.

Rancière, J. (1999 [2004]). *Disagreement: Politics and Philosophy*. Minneapolis, MN: University of Minnesota Press.

Rancière, J. (2006). *Hatred of Democracy*. New York: Verso.

Rawls, J. (1999). *A Theory of Justice*. Cambridge, MA: Belknap Press.

Sloterdijk, P. (2011). *Bubbles: Spheres Volume I: Microspherology*. Los Angeles, CA: Semiotext.

Swyngedouw, E. (2010). Impossible sustainability and the post-political condition. In Cerreta, M., Concilio, G., and Monno, V. (eds), *Making Strategies in Spatial Planning*. Dordrecht: Springer, pp. 185–205.

Torfing, J. (1999). *New Theories of Discourse: Laclau, Mouffe and Zizek*. Oxford: Wiley-Blackwell.

Urban, W. (2008). Who needs Yalom when we have Žižek?. *International Journal of Zizek Studies*, 2(2): 1–22.

Wigley, T. M. L. (2006). A combined mitigation/geoengineering approach to climate stabilization. *Science*, 314(5798): 452–454.

Yiftachel, O. and Hedgcock, D. (1993). Urban social sustainability: The planning of an Australian city. *Cities*, 10(2): 139–157.

Žižek, S. (1989 [2009]). *The Sublime Object of Ideology*. London: Verso.

Žižek, S. (2004). The structure of domination today: A Lacanian view. *Studies in East European Thought*, 56(4): 383–403.

Žižek, S. (2011). *Living in the End Times*. New York: Verso.

3 Urban social sustainability policies in the Nordic region

A repackaging of the welfare state model?

Moa Tunström

Introduction

In this chapter, I discuss urban policy and its contribution to social sustainability in Nordic cities while simultaneously recognising two difficulties in this endeavour. First, what is a socially sustainable city or neighbourhood? The concept of urban social sustainability is made meaningful in many ways in the contemporary era. It might be a city where people live long and healthy lives or a neighbourhood where the residents are organised and engaged in local politics. It might imply efforts to achieve a *socially mixed* neighbourhood, or measures to reduce socioeconomic differences. All these examples can be equally valid as social sustainability measures or ideals. Second, when making statements about social sustainability in the Nordic context, it is necessary to make generalisations. The 'Nordics' consists of countries and territories where policies and practices related to social sustainability vary, as does the understanding and actual use of the social sustainability concept. Nevertheless, in relation to the historically acclaimed model of a Nordic welfare state, understanding social sustainability in this broad context is still relevant (Christiansen and Markkola, 2006; Jørgensen and Ærø, 2008).

The background to this chapter includes my collected experiences and reflections from a number of Nordic research projects with which I have been involved in different capacities, all dealing with aspects of urban social sustainability, e.g. by investigating forms of citizen participation in Swedish urban planning (ongoing research), collecting segregation research (Tunström *et al.*, 2016) or examples of local integration initiatives (Nordic Welfare Centre, 2018), or identifying the role of civil society in Nordic post-war planning (Borges *et al.*, 2017). Working on these projects has forced me to reflect both on how the concept of social sustainability is defined and understood in general, and on the similarities between the Nordic countries. What is specific for the Nordics? These experiences have added to my understanding of urban social sustainability as a policy *discourse* and a framework for particular narratives, images and constructions of both problems and solutions in urban policy. This chapter is not an evaluation of the effectiveness of individual policies, but rather a reflection on the linkage between existing policies in the Nordic region and certain constructions of social sustainability.

Nordic cities are segregated and those who suffer the most from structural discrimination and crowding on the housing and labour market are immigrants from outside the EU. In this sense, many cities are far from socially sustainable (Tunström *et al.*, 2016). Even in the small country of Iceland, ethnic segregation is becoming a topic of both research and policy (Sindradóttir and Júlíusdóttir, 2008). Urban segregation presents a profound challenge and perhaps a particular challenge in relation to the self-perception and ideals of the Nordics that are based on the inclusive welfare state to a high degree. The purpose of this chapter is to discuss social sustainability measures in urban policy in the context of the Nordic welfare state. What types of measures and constructions of social sustainability are common? In the current situation, with increased privatisation and marketisation of previously public services, what do these dominant constructions of urban social sustainability tell us about continuations of, and breaks with, the historically strong welfare state? This chapter introduces these questions, approaching social sustainability as a discourse rather than as a distinct policy area – making generalisations, yes, but also hopefully spurring further debate on Nordic differences.

This chapter introduces the fuzzy concept of 'social sustainability': first as it is defined and used in research and policy, and thereafter in the specific Nordic policy context. Following this, three different conceptualisations of social sustainability in the Nordics are presented, and the chapter ends with a concluding discussion.

Social sustainability is a buzzword

The social aspects of sustainability have climbed the policy agenda during the last decade and are very much at the centre of the sustainability discourse in planning, both in the Nordics and elsewhere (see, for example, Boverket, 2010; Manzi *et al.*, 2010). Research calls and seminars have explicitly and implicitly emphasised social sustainability in both local and national policy in the Nordic context. Although the sustainability discourse was once mostly about new environmental solutions or reduction of carbon dioxide, it has come to be more and more about people and policies, i.e. about citizens, their lifestyles, political change and social justice. This has also resulted in social sustainability becoming somewhat of a buzzword. The term is used to signal the need to discuss environmental sustainability in the context of people and politics, but it is also used as something separate from 'the environment', something strictly about social interactions, social networks, etc. (Manzi *et al.*, 2010; Tunström *et al.*, 2015). Social sustainability in the urban context often covers a category of planning and development initiatives with labels such as 'integration', 'neighbourhood development' or 'local democracy', depending on the specific case. This chapter is concerned primarily with the Nordic expression of these initiatives and on broader urban policy-related conceptualisations rather than specifics such as education or employment.

Social sustainability initiatives are part of a diverse range of policies and practices with 'fuzzy' definitions (Vallance *et al.*, 2011). Some people claim that the social sustainability discussion should be about the basic needs of weaker groups, whereas others emphasise the need for change in the behaviour of those with resources (Vallance *et al.*, 2011). In contemporary Nordic urban development, it is possible to relate these two diverging tracks to a problem-oriented narrative about segregation and social injustices in existing neighbourhoods, and to a positive narrative about the appropriate preconditions for building community in new developments. Even though it can be hard to identify the relationship between, for example, area-based segregation measures in 'problem areas' and the marketing of well-designed parks and public spaces in new-built areas, contemporary urban discussion should be able to focus on both conceptualisations of social sustainability.

In the introduction to an edited volume on urban social sustainability, Manzi *et al.* (2010) present several ways to understand the concept. According to them it can concern social inclusion and exclusion, feeling at home in the neighbourhood, citizen participation, empowerment and democratic governance, and the importance of integrated planning of housing, services, transportation, health, education, etc. From the Nordic perspective, what they describe is actually urban planning in the Nordic countries during the build-up of the welfare state in the 1940s to 1970s. I return to this point later in the chapter. Manzi *et al.* (2010: 14) further observed that '[s]eeking social sustainability promotes new alliances and new ways of mobilizing resources through the new machineries of governance'. New alliances imply new ways of solving problems that have traditionally been the responsibility of established government structures. Increased citizen participation in urban planning and development is a clear example of this. Involving new groups and finding new ways of organising the planning process are seen as a remedy to the ills of previous eras.

Boström (2012) emphasised accessibility, participation, and social indicators and measurements. He stressed the importance of an awareness that different social groups have different resources and preconditions to access basic services, that several actors need to be involved in a socially sustainable process, and finally the importance of being able to measure 'the social' to see the possible imbalances in a place, neighbourhood or city. What this primarily indicates is, again, that social sustainability is a fuzzy concept (Allmendinger and Haughton, 2012; Markusen, 1999), and that finding a single definition that can work over time and place is, most probably, impossible. According to Boström (2012), social aspects of sustainability are more local, ideological and political, and harder to pin down and legitimise, than ecological aspects. Also, even though sustainability indicates sustaining something, social aspects are primarily about development and change, rather than preserving something in its current form. Consequently, this fuzzy concept has the potential to become the primary window, or looking glass, through which contemporary societal development can be understood and analysed. It can function as a spotlight, shedding light on power relations and inequalities, on social norms, and on the importance of social relations and networks. In that sense, it is a political concept. It is also political in the sense that it can be contested (Murphy, 2012). The common denominator of many definitions and indicators of social sustainability is that they are, or perhaps they should be, about social and spatial justice (e.g. Fainstein, 2010; Murphy, 2012). This implies just access to commercial and public services, to affordable housing, and to the opportunity for all citizens to participate in planning processes.

This chapter describes different understandings of urban social sustainability that are apparent in the Nordic discourse. A strict definition of the concept is not stressed. Instead, the different uses of the term and the associated political practices here indicate its meaning and importance in different contexts. In addition, contemporary social sustainability discourse in the Nordics must be understood through its relationship to the Nordic welfare state.

Planning a Nordic welfare state

The Nordic countries and cities are similar in many respects. From a European perspective, the Nordic region is remote and sparsely populated. The countries are welfare states with urban geographies that are dominated by small and medium-sized cities, which have all experienced relatively recent urbanisation. Today they have similar patterns of urban segregation, which has become a subject of debate along with concerns about socioeconomic inequality, welfare provision and immigration. These patterns involve cities with exclusive inner areas and low status housing areas – the 'segregated' areas – in the urban periphery, a segmented housing market and a strong ethnic component in the segregation patterns, especially in Sweden, Denmark and Norway (Tunström *et al.*, 2016). Furthermore, in each of the Swedish, Norwegian and Finnish languages, there is a similar term for planning that refers to more than just land use or the built environment. These terms (in Swedish *samhällsplanering*, in Norwegian *samfunnsplanlegging* and in Finnish *yhdyskuntasuunnittelu*) are similar combinations of the words for *society* and *planning*, and they all consequently refer to planning as the planning of *society* in physical, economic, social, political and cultural terms. Despite this broad meaning, these words also correspond to 'urban and regional planning' or 'town and country planning' as commonly used in international discourse. They can even refer to 'spatial planning' (see, for example, Schmitt and Smas, 2018). The fact that it is possible, and very common, to talk about planning in a broad sense immediately gives it significance beyond the built environment, i.e. it bridges the gap between town and country or urban and regional levels. As Schmitt and Smas (2018) also emphasise, the Nordic languages also strengthen social aspects of planning by pointing to the need to include aspects that are not explicitly related to land use or built environment.

Planning the city means planning for social interaction, housing for all, public transportation and accessibility to services, at least conceptually. In a Swedish context, Denvall (1997) emphasised that the term 'planning of society' must be understood as political and ideological, and that it must be placed in the context of the Nordic welfare state. Even though all planning is, in some sense, social, the social dimension has been further strengthened by the strong critique of 1960s' modernist planning (Denvall, 1997).

The Nordic welfare state, as a concept and as part of international discourse, was established in the 1980s, but before that the idea of a Scandinavian, or social democratic, model filled a similar function (Nygård, 2013). Even though there are differences between the Nordic countries, the concept has an important discursive function. Socioeconomic reforms have derived from the norms and values that even today are used in Nordic branding campaigns, i.e. trust, equality and democracy (Borges *et al.*, 2017). The Danish sociologist Gøsta Esping Andersen (1990), who first formulated this idea, considered high social spending, generous social policy and long social democratic dominance in politics as key expressions of the concept. In the literature, the Nordic welfare state model is usually defined as universalist, based on redistributive social policy and high public spending, and characterised by expert rule and 'social engineering' (Christiansen and Markkola, 2006; Denvall, 1997; Jørgensen and Ærø, 2008; Nygård, 2013). This social engineering concerns the organisation of family life, working life and the home. A further important characteristic is that the social security system is organised as a relationship between the state and the individual, and not based on affiliations with working life (Christiansen and Markkola, 2006). There have been paternalist traits to the welfare state model, such as the idea of a strong state taking care of and disciplining its citizens. In the post-war era, the focus shifted somewhat from the family to the individual, which allowed more women to enter the labour market, men (in varying degrees) to take paternity leave, and the nuclear family to not be the only possible social unit encouraged by policy (Christiansen and Markkola, 2006, Nygård, 2013).

The planning and development of housing and neighbourhoods in Nordic cities from the 1940s to the 1970s is a key manifestation of the *planning of society* and the welfare state. It might be said that the neighbourhood planning of that era influences today's social sustainability policy, both in the sense that ideals of the earlier era still shape the content of contemporary policy and because of the need to solve problems arising from modernist planning ideals. At that time, cities grew through development on previously undeveloped land at the urban periphery. New suburbs were characterised by accessible local services and activities, and a family orientation in both the design of the homes and their surroundings. Thus, community spaces and the dominance of public housing made a huge impact on the standard of urban life. High public spending resulted in high standards for housing, public transport and public spaces. Increasing female participation in the labour market resulted in public kindergartens and schools. Also important were the housing *policy* elements of these types of developments. The municipalities (Sweden and Norway) or the unions (Denmark) have been the main public or social housing providers and tenants have held a comparatively strong position (Bengtsson, 2013a, 2013b). Bengtsson (2013a, 2013b) clearly illustrated the difficulty in making generalised claims about Nordic housing policy (see also Jensen, 2013). Nevertheless, it is possible to claim that there has been a general 'right to housing' orientation, and that this can be understood as reflecting socially aware housing policies.

Since the 1940s, Nordic planning has traditionally been comprehensive, regulatory, rational and hierarchical, but has also recently developed more flexible strategic and market-oriented practices, as well as emphasising collaboration between public and private actors (Smas and Fredricsson, 2015). An underlying hypothesis in this chapter is that the Nordic welfare state, or at least the heritage and traces of the Nordic welfare state model, is responsible for the similarity of developments in contemporary Sweden, Norway, Denmark, Finland and Iceland. However, because of policy shifts towards increased privatisation and marketisation of previously public housing, education or health-care services, the Nordic welfare state is not considered to be as strong and universal as it was (Jørgensen and Ærø, 2008; Nygård, 2013; Szebehely and Meagher, 2013). This shift has been expressed in policies that affect not only the inclusiveness of social sectors, but also the built environment of Nordic cities. In Sweden, a lot of public housing was privatised during the 1990s and 2000s (Hedin *et al.*, 2012), and other examples of a transformed welfare state are decreased public social spending, stricter needs testing, raised pension age and investment of pension savings in the stock market (Nygård, 2013). Nygård (2013) calls this a politico-economic transformation and a turn away from the welfare state, which has led to both increased economic liberalisation and larger income disparities.

Civil society is an important actor

Civil society has an important role, both historically and currently, in the implementation of the welfare state model. Even though the Nordic welfare state model is based on a strong state, active public governance and civil society, in the form of associations and non-governmental organisations (NGOs), has had a vital role. In the Danish context, Ibsen and Habermann (2005: 6, my emphasis) observed that: 'The voluntary sector personifies personal freedom, social community and the fight for human rights, is the fertile soil for self-help, empowerment and integration, and functions as a *school for democracy*'. Historically, voluntary organisations, charities, foundations and grass-roots organisations have been key to outcomes that today are often taken for granted, such as organised public participation in local development, public housing and social welfare. In that sense, civil society associations can be considered important in teaching people about democracy, just as Ibsen and Habermann (2005) claimed. In addition, the historical focus on issues such as temperance, workers' rights and religious freedom is of importance to the conception of social sustainability. The activities of civil society organisations now also include, for example, women's shelters, environmental education, sports, local urban development, urban gardening or language training for immigrants. As the planning system has developed into a mix of public and private logics, awareness of the changing role of civil society adds to our understanding of the specific meaning of social sustainability in Nordic rhetoric and practice today. For example, the somewhat hesitant role of activism in the Nordics can be understood as a result of trust in government and democratic organisations, and also of the work of collectively organised interests in emphasising equality and inclusion, as well as social justice. In the Nordics, civil society is primarily not founded on activism or protest organisations, but is seen as representative of the people and as a dialogue partner to the state (Kings, 2011).

In the rest of this chapter the focus is on contemporary Nordic urban social sustainability discourse. I make clear that social sustainability in the Nordics is primarily a government project with traits derived from the welfare state, rather than a grass-roots initiative.

Conceptualisations of social sustainability in the Nordics

How is social sustainability made meaningful in urban policy and practice in the Nordic context? In this section, a few central aspects related to local democracy and participatory planning, community and the neighbourhood scale, and social mix policy in housing are discussed. These three themes are not uniquely Nordic, but their specific Nordic applications do illustrate the status of the Nordic welfare state. They exemplify what social engineering and the *planning of society* used to be, and of how contemporary social sustainability problem are expressed in policy. As a parallel, the American urban researcher Susan Fainstein (2010), in her conceptualisation of the just city, emphasises the opportunity for all citizens to participate in planning processes, just access to commercial and public services, and affordable housing as key components. This means, for example, to avoid gentrification by supplying affordable housing, to balance commercial activities with public services, and to establish low physical, economic and social barriers through services including public transport and generous public spaces.

Local democracy and a strong participatory discourse

Nordic planning is characterised to a high degree by the importance of local considerations. Municipal authorities and local politicians are the main decision makers when it comes to planning and, despite the existence of public–private partnerships and the involvement of consultants in the making and execution of plans, planning is still considered primarily a public responsibility. The municipality has long been the central service provider and urban planner. In Sweden, it is referred to as the municipal 'planning monopoly' and is often highlighted as a quality. Moreover, in all Nordic countries planning and building legislation provides opportunities for the public to express their views on ongoing plans. A rather strong participatory discourse is thereby already built into the system and is practised through tools such as open information meetings, public exhibition of planning proposals and the possibility for stakeholders to appeal the plans (Borges *et al.*, 2017; Fredricsson and Smas, 2013). In addition, via large member organisations and NGOs, civil society is an established dialogue partner with the state in the development of policies and in urban planning. These organisations have a central role as representatives of citizens or as providers of forums for dialogue. Groups, such as trade unions and tenants' organisations, are often complementary to the government or to private services, in the sense that they can organise and campaign for interests or mediate the relationship between the state and its citizens. NGOs are normally invited to comment on planning proposals. In Sweden, the Swedish Union of Tenants (see www.hyresgastforeningen.se/in-other-languages/engelska) also negotiate rents and terms and conditions of housing with landlords, as well as lobbying for good conditions in rental housing (Bengtsson, 2013b). Civil society forums and activities are focused on building social capital and safeguarding local democracy; therefore, they naturally carry promises, particularly relating to the social aspects of sustainability. They participate in making the citizen into an actor with local expert knowledge for the governance system to include and utilise.

Despite the strong participatory discourse in Nordic planning – formalised participation through public hearings, invitations to organisations to comment on proposals and tools ensuring that the 'public interest' is considered in new development – there is a strong push to further enhance the democracy, legitimacy and efficiency of the urban planning process (e.g. Kugelberg and Trovalla, 2015; Lindholm *et al.*, 2015). For example, the Swedish Association of Local Authorities and Regions (SALAR) provides national networks for knowledge exchange on new methods and tools for planning participation within Sweden. They collect and disseminate, via their website, examples of the use of participatory methods in urban planning (SALAR, 2018). They also conduct seminars and projects and encourage experimentation with emerging opportunities such as participatory budgeting and digital tools. Other examples of more experimental participatory methods that are gaining currency include 'urban labs' where public and private actors, and citizens, come together to develop proposals for a particular site (e.g. applications of this method can be found in the Baltic Urban Lab project – see www.balticurbanlab.eu) and e-governance-oriented activities in which people are invited to participate via web surveys, interactive maps or computer games.

Local authorities may issue the invitations for this kind of participation, but it is often private consultants who carry out the participatory events, consultants who can provide both technological innovations and attractive communicative tools and methods. This kind of experiment seeks to strengthen the democracy of the urban planning process by inviting novel forms of participation and attracting participation from new groups, such as young people, immigrants or those who usually do not speak up at traditional public meetings. However, they can also be considered as mere communication activities designed to strengthen the legitimacy of urban planning policy. Examples of this kind can be found in a local planning context throughout the Nordics. Not only municipalities, but also developers, architects and housing companies include both communicative and participatory activities in their planning processes. The methodological experimentation, the involvement of private consultants and the encouragement of using new methods to reach new groups, all seem to indicate the limitations of the planning systems. In ongoing research on the form and impact of participatory planning in Swedish municipalities, we are, among other things, interested in when dialogue became the key planning tool that it is seen as today, and what this really means for local democracy. Are all the focus groups, walking tours, open space meetings and events in the local shopping centre reaching new groups? Are the results from these participatory experiments properly integrated into the formal planning process? Dialogue is, like social sustainability, another 'fuzzy concept' used both formally and informally, and dialogue practices are not documented in any strict fashion. As a result of this, the collection of empirical material on participatory experiments in urban planning means searching for, and perhaps even constructing, the dialogue. The high expectations that come with participatory practices and their designation as *dialogues* can also be risky, in the sense that their inherent vagueness results in ambiguous policies and also in tensions between the established planning system and the initiators of the experimental dialogues. The participatory practices risk becoming merely a side activity to the formalised planning process. There are already indications that participatory planning, as practised, still means only reaching the information and consultation steps on Arnstein's (1969) ladder of participation (Gilljam and Jodal, 2005; Listerborn, 2007; Stenberg *et al.*, 2013). The participatory aspects of dialogue consequently need to be scrutinised and perhaps even questioned. There is a risk that the new partnerships and governance networks that often come into being because of dialogues – participatory initiatives and social sustainability projects – can blur the boundaries and divisions of responsibilities across public, private and community actors (Manzi *et al.*, 2010). Instead of overcoming the hierarchies, they are hidden. Another critical point in relation to participation is that it can 'localise' issues too much and come at the cost of sustainability. Australian planning researcher Leonie Sandercock's (1998: 190) statement that 'community' as an ideal, and the search for community through urban planning, has reached a 'semi-religious aura which seems to have the capacity both to inspire and to polarise any group of planners or planning students' remains a valid point. In the Nordic countries, there is a strong belief in the local community as place and in participatory methods as planning tools. The local community as the location for social sustainability measures, instead of the city as a whole, or society in an even broader sense, has resulted in localised practice. This is the focus of the next section.

The neighbourhood as a place and social network

The emphasis on local democracy is based on the understanding that it is at the local level that planning decisions should be taken and where people's everyday lives can be improved. It is at the local level that citizens travel by public transport, play with their children in a park or go grocery shopping. It is there that they may become friends with their neighbours or join organised activities in their spare time. All this indicates the importance of the local level, the neighbourhood and everyday life to social sustainability discourse and practice. The neighbourhood is a production site for *social capital* and is a product of the *planning of society* on a smaller scale. In addition, the urban design of a neighbourhood can favour segregation or integration. This understanding implies other ways to practise socially sustainable urban development apart from a participatory planning system, as discussed above. Importance is then placed on measures such as public health, on urban design or on responding to children's perspectives in the neighbourhood design.

Again, there is a need to generalise here in identifying what the Nordic countries have in common. In general, the importance of the local neighbourhood, its amenities and public spaces can be seen in the overall relatively well-designed and open Nordic cities. Parks and green spaces are predominantly public spaces and there is high awareness of the importance of accessibility for people with disabilities. This is an obvious but important implementation of a social sustainability perspective, i.e. to not assume that all citizens are equal and consider this in the design of public space. The Nordic countries have agreed on a charter for universal design (Björk, 2014), i.e. to use design as a tool for inclusion of elderly people, children and people with disabilities. In the planning process, the perspectives of the same groups on managing and feeling safe in urban public spaces is already emphasised through formal checkpoints in the planning process with regard to the consequences of planning proposals specifically for these groups, as well as through open public consultations, to ensure that these 'public interests' are considered in the design of physical environments and their amenities. At the very least, the system is organised to make this possible. Naturally, this emphasises social sustainability as a very local issue and a design issue. Social justice is constructed in terms of specific aspects, such as just access to local services and the public spaces of the city.

Inclusion is a strong aspect of *planning systems* in the Nordics. For example, it emerges from the influence of civil society on the planning process that was discussed in the previous section. It is also a consequence of a particular understanding of urban space, originating from the neighbourhood community planning ideals of the 1940s to the 1970s. Famous examples are Årsta and Vällingby in Stockholm, Sweden, and Tapiola in Espoo, Finland. To claim that housing areas from this era cause the values of the democratic welfare state to materialise is not particularly controversial, and it is possible to see this heritage as important in the contemporary construction of social sustainability ideals, even though the policy context has changed quite dramatically. However, traces of historical planning ambitions remain visible. The ambition to develop small-scale neighbourhoods with good standards of housing, services, public transport accessibility, and access to open parks and green spaces was not described as social sustainability because this concept was not part of the discourse at the time. The vision was not only accessible housing for everyone and a functioning local everyday life, but also their being accompanied by ambitions to influence people's values and the way they interacted socially. In an influential Swedish government investigation on housing and social policy from 1945, the architect and planner Uno Åhrén (1981) explicitly stated that this neighbourhood planning and design was supposed to create a 'democratic human being'. A key factor in the realisation of this intention was the small scale, but also spaces such as community halls for meetings, evening classes and cultural events, to encourage social interaction, education and democratic values.

Another way of seeing the importance of the local neighbourhood as the unit where social sustainability is to be achieved is the strong influence that *area-based initiatives* have had on the practice of reducing urban segregation. This is an example of a more problem-solving-oriented approach to social sustainability. It has been the mainstream strategy for some time, but only recently explicitly framed as 'social sustainability' (Tunström *et al.*, 2016). The implemented measures in an area-based programme can be diverse and oriented to the built environment, cultural events and education, but the main issue of importance in this chapter is how initiatives of this type locate both the social *problem of* segregation and its *solution* within the local area. The measures signal that it is in the 'segregated' neighbourhood that attractiveness needs to increase, where jobs should be created or people should be more politically engaged. It is in the neighbourhood that you can find the reasons for low voting or bad public health. There is scepticism and critique among researchers about the effectiveness of such approaches (e.g. Andersen, 2002). The sceptics claim that area-based initiatives can improve local conditions, but they cannot solve the problem of segregation and socioeconomic divisions in the city as a whole. In the Swedish context, recently, the so-called 'social sustainability commissions' that have been initiated by cities are signs that policy is moving away from the area-based approach to amelioration of segregation and social injustices, and instead taking on a broader and city-wide approach. At the time of writing, the commissions in the cities of Malmö, Gothenburg and Stockholm are the most well known. Instead of localising problems and solutions only to the 'segregated' areas, the commissions start from the segregated *city* and map socioeconomic differences city wide, which makes it possible to approach issues of what is being built where, how the city is governed, and what a city-wide mapping of public health, education level and other indicators reveals about social and spatial injustices. However, there is simultaneously a strong Danish policy discourse where certain housing areas are officially designated as 'ghettos', which points in the completely opposite direction by clearly differentiating those housing areas and their citizens from the city as a whole (Bakkær Simonsen, 2016; Baskerville *et al.*, 2011).

Mixed housing is key

A common strategy for reducing socioeconomic segregation and increasing social sustainability is development of policies aimed at increasing the social mix. Social mix as a concept can be used as an urban planning theory term, or it can refer to socioeconomic composition, or to certain urban transformation strategies (Galster and Friedrichs, 2015). Galster and Friedrichs (2015) further point to the political ambiguity of the concept, in that it has been regarded as both a neoliberal and a socialist concept. Social mix policies are intended to facilitate a mixed composition of residents in an area, which results in diversity and variation with respect to residents' socioeconomic resources (Christensen, 2015; Holmqvist and Bergsten, 2009). This is considered to strengthen the social sustainability of a neighbourhood or a city, which could be understood as a more social justice-oriented agenda, if it means increasing the opportunities for groups lacking resources, or a gentrification agenda, if it means the opposite, e.g. regenerating an area to attract those with resources. The type of social mix that is referred to also varies from a mix of residents with distinct socioeconomic resources to an ethnic mix, household types or a mixed housing tenure (Galster and Friedrichs, 2015). Consequently, the effects of a social mix strategy are not yet clear.

When social mix is on the agenda in Nordic countries, it usually refers to mixing *housing types and tenure forms* to attract or locate different social groups and lifestyles to the same city, district, housing area or block. In addition, the issue of whether to own or rent a dwelling is considered key to the mixing of socioeconomic groups, ethnic groups and lifestyles (Holmqvist and Bergsten, 2009), and it is thought to be vital that individuals and groups that differ in these respects can meet in an everyday life setting. In general, there is a shortage of affordable housing in the Nordic capitals (Tunström *et al.*, 2016). Whether or not a city has social housing, as well as the general status of rental housing, is then of considerable significance. Social mix policy in the Nordic countries is consequently primarily a socioeconomic and housing policy-related issue, and less related to ethnicity. Another difference compared with other countries is that it has rarely been about demolition of rental housing to make way for other tenure forms, but instead oriented to new construction (see, for example, Holmqvist and Bergsten, 2009). Finland, and particularly Helsinki, stand out in the Nordic context when it comes to social mix. In Helsinki, the goal has for several decades been that around 20% of all newly developed housing, regardless of where in the city, should be state-supported rental housing with a share of housing categories for special groups: around 40% should be of the mixed and regulated forms of so called 'part-ownership housing', 'right-of-occupancy housing' and regulated ownership housing (so-called 'Hitas housing'), and around 40% should be unregulated rental and ownership housing (City of Helsingfors, 2012, 2016). The Helsinki municipality housing and land-use programme is quite explicit in its aim to build good quality housing for different social groups and different tenure forms. In the Nordic context, the Finnish model stands out also because it orients social mixing to the block level (City of Helsingfors, 2012, 2016). Compared with Finland, Swedish policy uses social mix on a less detailed scale and thereby with a less obvious effect, even though this has been an element in housing policy since the 1970s (Holmqvist and Bergsten, 2009). The housing market is also not as diverse, with basically only cooperative ownership or rental housing as the available tenure forms in new developments. However, in both the Finnish and Swedish examples, it should be noted that what can be achieved is a socially mixed group of residents in a block, neighbourhood or city, but no guarantee of integration in the sense of social interaction, or in the sense of empowering or even strengthening those who are in a weak socioeconomic position.

Conclusions: a contemporary urban policy discourse that can be traced back to the Nordic welfare state

The power of social sustainability as a concept and policy complex is that it can unite several different objectives of contemporary urban development in the areas of planning process and governance, urban design and social initiatives. It is a strong idea that invokes a lively discourse, but it is not yet strongly tied to particular practices. Even though I have tried to connect contemporary social sustainability thinking to the *planning of society* and the Nordic welfare state in this chapter, it is clear that, although the welfare state induced a built form that expressed the neighbourhood planning ideal, a material form arising from the social sustainability ideal remains to be seen. In their comparison of 'urban policy' in the Nordic countries, Jørgensen and Ærø (2008: 37) observed that 'urban policy does to some extent conflict with the aim of the welfare state to help the weakest the most.... Many resources, both money and attention from the state, are spent in relatively few project areas'. If the welfare state model was 'unitarist, egalitarian and based on the active participation of residents and local actors', urban policy is – and I claim similar goals for social sustainability policies – 'spatially and socially targeted and dependent on active participation of residents and local actors' (Jørgensen and Ærø, 2008: 25).

The fact that social sustainability is a positive concept and that it directs our attention to solutions instead of problems makes it powerful. However, this does not always mean that it leads to the realisation of something radically new. By investigating the manifestation of urban social sustainability discourse in the Nordic context through lenses such as a local development perspective, a participatory perspective and mixed housing tenures, social sustainability appears as a continuation of a quite traditional perspective. It is a way of understanding planning as being about the *planning of society*, and thereby includes issues about democracy and community, in addition to the physical planning and design of infrastructure and the built environment. At the same time, the fact that the term 'social sustainability' has climbed the policy agenda during a period of fragmentation of the Nordic welfare state, through privatisation of public services, a weakening public sector and increased socio-economic differences in Nordic cities, could be a sign that the relationship between urban planning and the welfare state has changed. The municipality is still a key actor, but their available tools are fewer than before. Participatory experiments could then be interpreted as piecemeal practices rather than improved formal participation, and mixing of tenure as strategic gentrification rather than as reducing the barriers to the housing market. The two versions of social mix policies from Sweden and Finland illustrate that it is difficult to claim something general about the effectiveness of contemporary policy; instead the outcome must be dependent on what tenure forms are available, and the scale on which the mix policy is implemented. Finally, it can be noted that, despite the positive and idealistic understanding of social sustainability, the practices undertaken in its name are largely about solving problems, e.g. solving segregation by mixing tenure forms, solving the lack of social capital by area-based community projects or solving the democratic deficit by participatory experiments. Less in focus are the social sustainability effects simultaneously arising in other parts of the segregated city in the form of exclusive new developments, or where deals are made between land owners and developers, which are processes that influence the socioeconomic divide within a city and that consequently create the preconditions for certain groups and certain urban lifestyles to emerge.

References

Allmendinger P and Haughton G (2012) Post-political spatial planning in England: A crisis of consensus? *Transactions of the Institute of British Geographers* 37(1): 89–103.

Andersen HS (2002) Can deprived housing areas be revitalised? Efforts against segregation and neighbourhood decay in Denmark and Europe. *Urban Studies* 39(4): 767–790.

Arnstein SR (1969) A ladder of citizen participation. *Journal of the American Planning Association* 35(4): 216–224.

Bakkær Simonsen K (2016) Ghetto–society–problem: A discourse analysis of nationalist othering. *Studies in Ethnicity and Nationalism* 16(1): 83–99.

Baskerville W, Jürgens N, Lord S, and Overton B (2011) In the ghetto or out of society? A review of the debate surrounding the 'ghetto plan'. Available at: www.humanityinaction.org/knowledgebase/385-in-the-ghetto-or-out-of-society-a-review-of-the-debate-surrounding-the-ghetto-plan (accessed 20 June 2018).

Bengtsson B (2013a) Varför så olika? Om en nordisk gåta och hur den kan lösas [Why so different? A Nordic riddle and how to solve it]. In: Bengtsson B, Annaniassen E, Jensen L, Ruonavaara H, and Sveinsson JR (eds), *Varför så olika? Nordisk bostadspolitik i jämförande historiskt ljus* [Why so different? Nordic housing policy in comparison]. Malmö: Égalité.

Bengtsson B (2013b) Fem länder, fem regimer, fem historier: en nordisk jämförelse [Five countries, five regimes, five histories: A Nordic comparison]. In: Bengtsson B, Annaniassen E, Jensen L, Ruonavaara H, and Sveinsson JR (eds), *Varför så olika? Nordisk bostadspolitik i jämförande historiskt ljus* [Why so different? Nordic housing policy in comparison]. Malmö: Égalité.

Björk E (2014) A Nordic charter for universal design. *Scandinavian Journal of Public Health* 42(1): 1–6.

Borges LA, Nilsson K, Tunström M, *et al.* (2017) *White Paper on Nordic Sustainable Cities*. Stockholm: Nordregio. Available at: www.nordregio.org/research/white-paper-on-nordic-sustainable-cities [accessed 20 November 2018].

Boström M (2012) A missing pillar? Challenges in theorizing and practicing social sustainability: introduction to the special issue. *Sustainability: Science, Practice & Policy* 8(1): 3–14.

Boverket (2010) *Socialt hållbar stadsutveckling: en kunskapsöversikt* [Socially sustainable urban development: an overview]. Karlskrona: Boverket.

Christensen G (2015) A Danish tale of why social mix is so difficult to increase. *Housing Studies* 30(2): 252–271.

Christiansen NF and Markkola P (2006) Introduction. In: Christiansen NF, Petersen K, Edling N, and Haave P (eds), *The Nordic Model of Welfare: A Historical Reappraisal*. Copenhagen: Museum Tusculanum Press.

City of Helsingfors (2012) *Hemstaden Helsingfors. Genomförandeprogram för boende och härmed sammanhängande markanvändning* [At home in Helsinki. Implementation programme for housing and land use]. Helsingfors: Helsingfors stad.

City of Helsingfors (2016) *Hemstaden Helsingfors. Genomförandeprogram för boende och härmed sammanhängande markanvändning* [At home in Helsinki. Implementation programme for housing and land use]. Helsingfors: Helsingfors stad.

Denvall V (1997) Inledning [Introduction]. In: *Välfärdens operatörer: Social planering i brytningstid* [The operators of welfare: Social planning in transition]. Umeå: Boréa.

Esping Andersen G (1990) *The Three Worlds of Welfare Capitalism*. Cambridge: Polity.

Fainstein S (2010) *The Just City*. Ithaca, NY: Cornell University Press.

Fredricsson C and Smas L (2013) *En granskning av Norges planeringssystem: Skandinavisk detaljplanering i ett internationellt perspektiv* [An investigation of the Norwegian planning system: Scandinavian detailed development planning in an international perspective]. Stockholm: Nordregio.

Galster GC and Friedrichs J (2015) The dialectic of neighborhood social mix: editors' introduction to the special issue. *Housing Studies* 30(2): 175–191.

Gilljam M and Jodal O (2005) *Demokratiutveckling i svenska kommuner del 3: Kommunala demokratisatsningar: vägen till en mer vital demokrati?* [Developing the democracy of Swedish municipalities part 3: Municipal democracy projects: the road to a vitalized democracy?] Göteborg: Centrum för forskning om offentlig sektor (CEFOS), Göteborgs universitet.

Hedin K, Clark E, Lundholm E, and Malmberg G (2012) Neoliberalization of housing in Sweden: Gentrification, filtering, and social polarization. *Annals of the Association of American Geographers* 102(2): 443–463.

Holmqvist E and Bergsten Z (2009) Swedish social mix policy: A general policy without an explicit ethnic focus. *Journal of Housing and the Built Environment* 24(4): 477–490.

Ibsen B and Habermann U (2005) *Defining the Nonprofit Sector: Denmark*. Working Papers of the Johns Hopkins Comparative Nonprofit Sector Project, No. 44. Baltimore, MA: The Johns Hopkins Center for Civil Society Studies.

Jensen L (2013) Housing welfare policies in Scandinavia: A comparative perspective on a transition era. *LHI Journal* 4(2): 133–144.

Jørgensen G and Ærø T (2008) Urban policy in the Nordic countries: National foci and strategies for implementation. *European Planning Studies* 16(1): 23–41.

Kings L (2011) *Till det lokalas försvar: civilsamhället i den urbana periferin* [In defence of the local: Civil society in the urban periphery]. Lund: Arkiv.

Kugelberg C and Trovalla E (2015) *Samverka och/eller styra: att praktisera medborgarinflytande i planering och beslutsfattande* [Collaborate and/or govern: Practicing citizen influence in planning and decision making]. Göteborg: Makadam.

Lindholm T, Oliveira e Costa S, and Wiberg S (2015) *Medborgardialog: demokrati eller dekoration? Tolv röster om dialogens problem och potential i samhällsplaneringen* [Citizen dialogue: Democracy or decoration? Twelve voices on the problem and potential of the dialogue in planning]. Stockholm: Arkus.

Listerborn C (2007) Who speaks? And who listens? The relationship between planners and women's participation in local planning in a multi-cultural urban environment. *GeoJournal* 70(1): 61–74.

Manzi T, Lucas K, Jones TL, and Allen J (2010) Understanding social sustainability: Key concepts and developments in theory and practice. In: Manzi T, Lucas K, Jones TL, and Allen J (eds), *Social Sustainability in Urban Areas: Communities, Connectivity and the Urban Fabric*. London: Earthscan.

Markusen A (1999) Fuzzy concepts, scanty evidence, policy distance: The case for rigour and policy relevance in critical regional studies. *Journal of Regional Studies* 33(9): 869–884.

Murphy K (2012) The social pillar of sustainable development: A literature review and framework for policy analysis. *Sustainability: Science, Practice, & Policy* 8(1): 15–29.

Nordic Welfare Centre (2018) Nordic co-operation on integration and inclusion. Available at: https://nordicwelfare.org/integration-norden/en/ (accessed 20 June 2018).

Nygård M (2013) *Socialpolitik i Norden: en introduktion* [Social policy in the Nordic region: An introduction]. Lund: Studentlitteratur.

SALAR (2018) Dialogguiden: ditt stöd för att planera dialoger [The dialogue guide: Support in the planning of dialogues]. Available at: www.dialogguiden.se (accessed 20 June 2018).

Sandercock L (1998) *Towards Cosmopolis: Planning for Multicultural Cities*. Chichester: John Wiley & Sons.

Schmitt P and Smas L (2018) Spatial planning in Sweden: Key notions, trajectories and challenges. In: Farinós J and Peiró E (eds), *Achieving Territory Becomes Matter of State Importance: Essentials for Coordination of Spatial Planning Policy*. Valencia: Tirant Humanidades, in press.

Sindradóttir JI and Júlíusdóttir M (2008) Búseta innfl ytjenda á höfuðborgarsvæðinu [Residency of immigrants in the capital area]. *Landabréfið – Tímarit Félags landfræðinga* 24: 3–22.

Smas L and Fredricsson C (2015) Changes and inherent tensions in the Nordic planning systems. *Nordregio News* 2: 4–7.

Stenberg J, Abrahamsson H, Benesch H, *et al.* (2013) *Framtiden är redan här: hur invånare kan bli medskapare i stadens utveckling* [The future is already here: How residents can become co-creators in the development of the city]. Göteborg: Chalmers tekniska högskola.

Szebehely M and Meagher G (2013) Four Nordic countries: Four responses to the international trend of marketisation. In: Szebehely M and Meagher G (eds), *Marketisation in Nordic Eldercare: A Research Report on Legislation, Oversight, Extent and Consequences*. Stockholm: Department of Social Work, Stockholm University.

Tunström M, Gunnarsson-Östling U, and Bradley K (2015) *Socioekologisk stadsutveckling: begrepp och lokal praktik* [*Socio-ecological urban development: concepts and local practice* [bilingual publication]. Stockholm: Arkitektur förlag.

Tunström M, Anderson T, and Perjo L (2016) Segregated cities and planning for social sustainability: A Nordic perspective. *Nordregio Working Paper* 2016: 3. Stockholm: Nordregio.

Vallance S, Perkins HC, and Dixon JE (2011) What is social sustainability? A clarification of concepts. *Geoforum* 42(3): 342–348.

Åhrén U (1981) *Ett planmässigt samhällsbyggande: särtryck ur bostadssociala utredningen* (SOU 1945:63) [*A plan-based urban development: offprint from the governmental housing investigation* 1945: 63]. Stockholm: KTH.

4 Social sustainability and transport

Making 'smart mobility' socially sustainable

Tanu Priya Uteng, Yamini Jain Singh, and Oddrun Helen Hagen

Social sustainability in the transport planning sector

The ethos of numbers and technocratic rationality overpowers other areas in transport-related development projects. Often expressed in 'kilometres of roads constructed', 'numbers of new highways built' in the past fiscal year, 'time savings' etc., the engineering and technical dimension routinely takes precedence. The essential denominator of development, i.e. the affected population, is given a backseat in a typical transport discourse and (aggregated) population ends up as mere numbers to be fed into the transport modelling exercises, typically projecting the need for capacity enhancement of road systems or mass transit systems. In this number-crunching exercise, only one pillar of sustainable development has historically been given priority – economic development. Increasing environmental awareness has led to altering this pattern by focusing on the goal of environmental protection along with economic efficiency. Despite this, the ways in which these 'calculated and projected' environmental and economic developments filter down to affect the general populace or the pillar of social sustainability remain largely ignored.

For example, sieving through published works by the Transportation Research Board on sustainable transportation till 2014 suggests that the three thematic areas of sustainability (economic, environmental, social) are not equally represented in the transportation literature. Even though the term 'sustainability' surfaced only in 1980s, and research on the three themes of sustainability actually started to appear after the 1990s, research focus on economic and environmental sustainability is visible. Lineburg (2016: 11) plots the occurrence of each sustainability sector in the transportation literature till 2014 and highlights that environmental sustainability has had more attention in recent times than economic and social sustainability because there was high awareness of environmental crisis brought about through the use of fossil fuels in the transport sector. As noted by Jeekel (2017), the attention on social issues, although considered important in reaching sustainable development goals, lagged behind in both practice and research arenas. The topic of social sustainability, therefore, demands further study through bringing together 'the relationships between individual actions and the created environment, or the interconnections between individual life-chances and institutional structures' (Jarvis *et al.*, 2001: 127).

Lineburg (2016) highlights the multiple overlaps embedded in the concept of social sustainability, and gathers them together under the conceptual umbrellas of 'social equity' and 'sustainability of community' as the two fundamental concepts of social sustainability in transportation. Although *social equity* includes accessibility, safety, and health, *sustainability of community* includes cohesion, participation, and awareness. Social sustainability is a complex subject and entails a wide array of topics that should ideally be linked through systems thinking and (evidence-based) impact analyses. Although there are disparate discussions on related topics, there have been few attempts to bundle and present them as the basis for discussing social sustainability. The two vital questions that need to be asked are:

1 How can social sustainability be understood in the transportation sector?
2 How can social sustainability be assessed while designing, implementing and assessing/monitoring transportation projects?

For example, the following topics are studied in separate branches of transport planning but a systemic structure to bring these topics together for comment and reflection on plans under discussion is largely missing:

- Mobility patterns of the different groups,
- Traffic accidents and safety issues,
- Affordability, for example pricing of public transport,
- Acceptability, for example decisions affecting livelihoods,
- Accessibility, for example route planning, timing,
- Health effects of transport,
- Planning for well-being and inclusive communities, etc.

Similarly, Lucas and Stanley (2013) mention that the problem in promoting this agenda is that currently there are no existing social sustainability criteria within the transport appraisal tools of major transport investors. We argue that this fallacy of a 'quasi' approach of analysing the social dimension of transport is present across the developed and developing world, even though the extent of the problem remains more pronounced in the developing world. Quantifying social sustainability and linking it to various development programmes therefore become crucial.

Going forward, we briefly highlight the methodologies that have been used in the past to assess social sustainability in the transport sector, and then use an approach, partially based on these methodologies, to examine smart mobility further. Currently, digitalisation and smart cities exert the biggest influence on the urban and transportation sector, necessitating discussion of future policy and research recommendations bearing in mind these smart elements. The chapter concludes with a brief discussion on policy and research recommendations.

Measuring social sustainability of transport projects

Social sustainability has been measured in multiple ways to date, and no standard methodology can be extracted easily and put out there. Measurements typically include a set of indicators and rating systems to assess social sustainability. Lucas *et al.* (2007) identify and validate a set of indicators to assess social sustainability of transport decisions in the UK. The aim of the overall project was to develop a tool that could be practically used by decision-makers (at both the national and the local levels) to make comprehensive assessments of the social sustainability impacts of transport policies and projects (Marsden *et al.*, 2007). The following five core indicators were developed:

1 Transport poverty: affordability of public transport relative to income for households below the poverty line;
2 Accessibility: weighted journey times to key centres of (a) employment, (b) primary, secondary, and further educational facilities, (c) primary healthcare providers and general hospital, and (d) key food shops;
3 Safety: adult and child pedestrian casualties by social class;
4 Quality of life: percentage of residents living within 1,000 m or a 15-min 'safe walk' to key destinations (e.g. health, educational, leisure and cultural facilities, food, shops, post office) by relevant social groups;
5 Housing availability: lowest 10% value of house prices within *x* minutes (based on average population local journey times to employment within any given location) of the town centre and key centres of employment.

In Lucas and Stanley (2013), the authors suggest further addition of the following five appraisal criteria to the list, which will also help in streamlining evidence-based analyses of social sustainability in transport discourse:

1 Health and well-being: including protection from road deaths and injuries, freedom from pollution and other adverse effects such as noise, major roads dissecting communities, traffic congestion, promotion of healthy travel and absence of social exclusion;
2 Equal opportunity to participate in society: including availability, accessibility, and cost of transport (for present and future generations and all sectors of the population), thus facilitating both social inclusion and equity goals;
3 Transparent and accountable transport governance structures: at every layer of decision-making, e.g. public enquiries, referenda and engagement exercises, community representation within transport organisations;

4 Access to decision-making processes and recourse to legal justice: including governance and planning structures, which allow bottom-up, community decision-making processes to occur, e.g. community-led local transport plans;
5 Integration: a cross-cutting governmental agenda for transport across multiple policy sectors, including housing, city planning, public utilities, health, education, environment, and social welfare, e.g. indicators of accessibility to key activities disaggregated by population groups, such as employment, education, and healthcare included within the policy performance measures for the different development sectors.

Another format of designing a practical check for the sustainability concept can be the adoption and standardisation of the *Transportation Rating Systems*. These were designed to quantify sustainability objectives through metrics, evaluation methods, and best practice models. The systems are based on a points-based certification of sustainable design and serve as a prominent methodology for developing assessment tools to evaluate the sustainability of transportation projects in the USA (Curz *et al.*, 2012). It allows users to incorporate sustainable practices into transportation processes and programmes by evaluating existing infrastructure and informing design decisions. But, given the paucity of standardised methodology, the extent to which the different rating systems analyse social equity, economic prosperity, and environmental health varies. Parallels exist between the published works by TRB and how these transportation frameworks disproportionately position and favour the evaluation of economic and environmental impacts over those of social concerns (Curz *et al.*, 2012; Dondero *et al.*, 2012; Litman, 2012). Drawing on the conclusions of Mercier (2009), Manaugh *et al.* (2015), and Litman (2012), Lineburg (2016) reinstates that the existing forms of incomplete analysis calls for a more comprehensive and systematic approach to defining, measuring, and evaluating social impacts to encourage incorporation of social sustainability objectives into transportation designs.

What is of interest for the purpose of this chapter are the topics covered to comment on social sustainability. These topics have been collated in Figure 4.1. As Figure 4.1 highlights, social sustainability can be broadly classified into the following two categories: *social equity* and *sustainability of communities*. These two respective categories can be further divided into subcategories, which branch out to cover the various domains falling under the social sustainability umbrella. Such a detailed classification and evaluation is vital given the multiplicity of topics involved when commenting on social sustainability. Dividing the evaluating criteria into 'process' and 'outcome' measures adds another layer to carve out a robust methodology for commenting and planning for social sustainability. The 'process measures' present a rather old and much-discussed approach to linking the fields of social development and urban/transport planning, but this kind of structure is important to highlight how the topic of 'transport and social sustainability' can be approached at the plan-making stages.

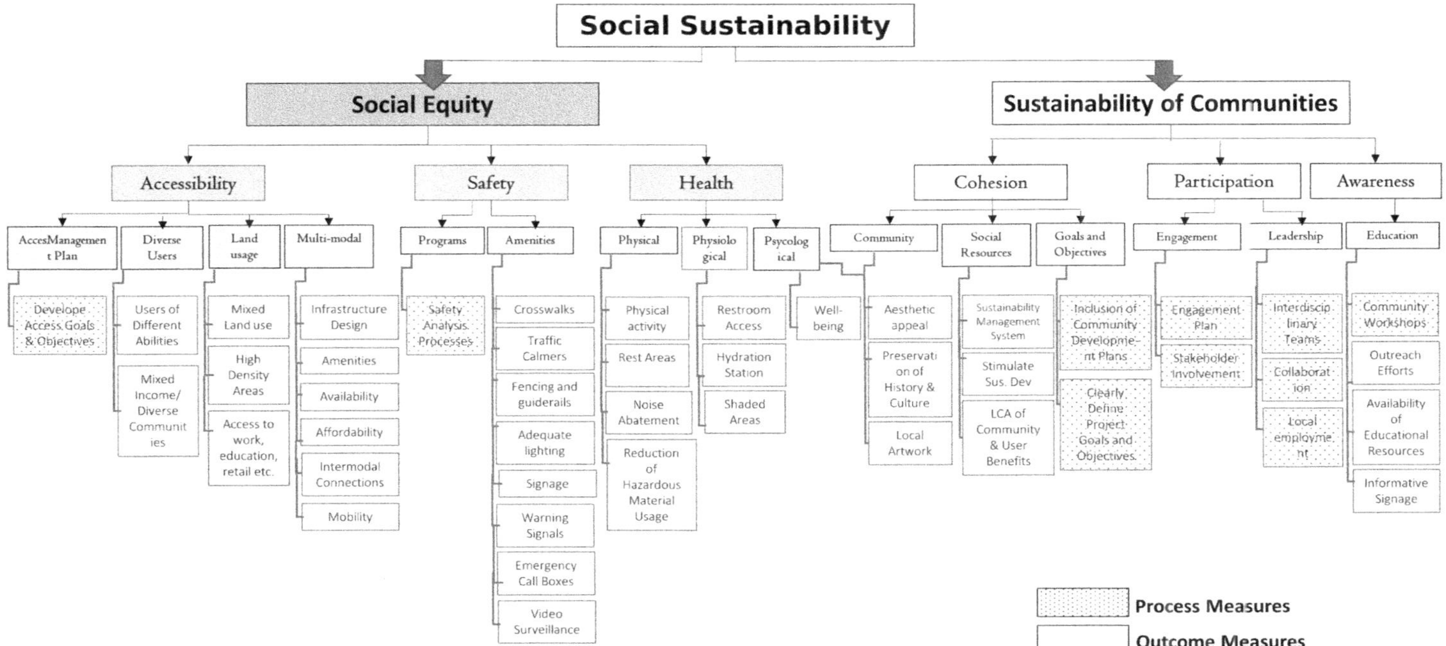

Figure 4.1 Elements of social sustainability.

Source: reproduced with permission from Lineburg (2016: 36).

The outcome measures presented in Figure 4.1 provide a comprehensive picture of elements that can be measured as outcomes of a particular transport project. Rather than being static, these measures should be seen as suggestive, adaptable, and dynamic in nature.

For example, the concept and advocacy of the '20-minute city' as a compact city where the lion's share of employment, recreation, and social activities are planned and located within 20 minutes from residential locations is fast gaining traction (Newman, 2016; Whitzman, 2017). Although the idea seems to be environmentally sustainable and inclusive, and promises to reduce disparities in access to services, it needs further evaluation from the standpoint of social viability and sustainability. Some of the 'outcome measures' that are relevant for assessment of a project such as the 20-minute city are: users of different abilities, mixed land use, high-density areas, infrastructure design, intermodal connections, crosswalks, traffic calmers, fencing and guiderails, adequate lighting, signage, warning signals, physical activity, and rest areas. After a prolonged interaction between the researchers, planners advocating the 20-minute city and the policy-makers, the Australian authorities have finally approved the concept plan for a 30-minute city. This goes on to highlight how the social sustainability structure can be applied to win the cause of designing sustainable cities built around the accessibility dimension.

Later, we describe an approach to assess smart mobility measures that have already been implemented. Our point of departure here is the credit assessment categories of Figure 4.1, and incorporating some of the 'outcome measures'. The 'outcome measures' were found to be more suitable for assessment of smart mobility in general, and some of these are used to structure the discussion concerning how (and if) social sustainability is being captured in the emerging domain of smart mobility. Commenting on 'process measures' would have entailed detailed document studies and interviews to investigate the steps taken before implementing smart mobility measures, a task that can be taken up in future research studies.

In the following sections, we delve into 'smart cities' and the smart mobility agenda to assess social sustainability based on the approach described. We discuss the extent to which smart mobility is socially sustainable. We also discuss how social sustainability in the transport sector can be inbuilt within the smart cities agenda right from the inception of the projects.

Assessing social sustainability of smart mobility

Smart cities and smart mobility

The scope and complexity of urban challenges have often been approached under different headings such as sustainable cities, happy cities, liveable cities, future cities. One of the conceptual umbrellas gaining ground as a winning strategy, this time with full governance and industrial backing, is that of smart cities. Smart cities, as an ideology and set of algorithms (to make all cities standard, be able to communicate with each other and monitor), are gaining wide support internationally. The term remains, however, as complex as the cities themselves.

The use of information and communication technologies (ICT) constitutes the core of the smart agenda. Although other sustainable approaches to city development involve mainly city planners and sectors traditionally intertwined with planning, high-tech companies, such as IBM, Siemens, and Bosch, have taken the front seat in designing the smart cities concept. And with a steady increase in actors involved in this concept, we also face new approaches and new definitions of smart cities understood differently by different people and sectors. Albino *et al.* (2015) list 23 definitions of the concept, and emphasise that smart cities cannot be broached through digital unpacking alone, but should essentially be examined to look at the ramifications for the environment, governance, and community needs. In other words, the smart city concept should not be limited exclusively to the dissemination of ICT, but also look at people and community needs. Enhancing the quality of life should become a goal of this approach (ibid).

Within the smart cities approach, smart mobility is a key component (Benevolo *et al.*, 2016) that aims to enhance effective mobility using information technology and digital data for rebooting the existing transportation systems. Smart mobility aims to make efficient use of the existing infrastructure and resources at a fraction of the cost required to build a new infrastructure and deliver the same benefits. Benevolo *et al.* (2016) have categorised the most important smart mobility objectives into six categories: (1) reducing pollution; (2) reducing traffic congestion; (3) increasing safety; (4) reducing noise pollution; (5) improving transfer speed; and (6) reducing transfer costs. Later, they stress that smart mobility involves all three smart city paradigms: digital city, green city, and knowledge city. Smart mobility solutions range from on-demand ride services, real-time ridesharing services, car- and bike-sharing programmes, multimodal trip-planning apps, and smart traffic control to self-driving vehicles.

Criticisms and limitations

Most of the criticisms about smart cities address the umbrella concept, but, as a central approach to the concept, these critiques also concern smart mobility. One of the criticisms of the smart city approach has been the lack of a common definition, clarifying which elements go into making up a smart city. Holland (2014) raises a critical concern about the involvement of corporations in the smart city concept, claiming that most smart city projects have a lack of concern with democratic decision-making and real citizen involvement. This may imply a change in the driving forces of city development, governance, and control in the future, which might prove to be a fundamental (and dangerous) shift from the very ethos of sustainable development. Holland (ibid) argues that these critical remarks raise important questions about the self-congratulatory nature of the smart city and how such ideas about this approach are being promulgated by its advocates and marketers. Docherty *et al.* (2017) find that smart mobility proponents sell smart mobility innovations as solutions for reducing wastage of productive time spent in congestion, improving well-being, etc., when in reality these are secondary impacts. For them, it is more important to grow their mobility market by tapping into the 'unmet lifestyle needs', which is in fact contrary to their claims that smart mobility aims eventually to reduce mobility needs.

Furthermore, another major criticism is the gap between the smart approach and quality-of-life approaches. Lauwers and Papa (2015) claim that this gap is found both in the literature and in practice, arguing that there is a paradigm shift from conventional mobility planning towards smart mobility, by applying new technology to existing infrastructure instead of creating better solutions. For example, buses are being retrofitted with tracking devices rather than increasing public transport supply and checking outcome measures such as access to work, education, etc. In this sense, smart mobility concerns itself primarily with innovative technological or consumer-centric solutions rather than adopting a social sustainability lens to the entire mobility agenda. The technocentric approach is characterised by ICT through management, control, and optimisation as its keystone. The consumer-centric approach is characterised by a strong emphasis on the demand side of transport systems and transport users as consumers of services (ibid.), but the calculation of demand itself is ridden with fallacies where work trips and car-based trips have been given the pivotal role.

Another criticism of the approach, which applies equally to the field of transportation, is the standardisation that 'smartness' will impose on cities, where common indicators would be applied for profiling cities, identifying areas of action for improving their smartness, and assessing the impact of actions on the city. Such indicators will be issued as a standardised methodology by August 2018 by the United Smart Cities Program[1] (jointly initiated by the United Nations Economic Commission for Europe and other industrial partners). But cities differ in their context, and standardisation also needs a certain level of flexibility to be adaptable to local conditions. The various available definitions of both smart cities and smart mobility underline that the approach lacks universality, and comparing cities might lead to a loss of information to deliver socially sustainable solutions in the transport sector.

Assessing social sustainability of smart mobility

In this section, we use the structure presented in Figure 4.1 to assess social sustainability in the smart mobility agenda. As mentioned earlier, social equity and sustainability of communities are the two main criteria within which there are 'process' and 'outcome' measures for assessment. As smart mobility solutions, such as on-demand ride services, real-time ridesharing services, car- and bike-sharing programmes, have already been implemented, the appraisal of such projects using 'process measures' is an extensive topic that needs a dedicated study and a longer time frame. On the other hand, it is relatively easy to comment on the 'outcome measures' to assess how socially sustainable these smart services are. The following sections build on a large body of literature that has studied the impacts of smart mobility services in different parts of the world.

Social equity

ACCESSIBILITY

In this section, we assess how accessible different smart mobility services to different sections of the society are. According to Litman (2000) and Duncan (2011), car sharing provides a more flexible option to owning a car because it comes with low fixed costs but higher variable costs. Users, who typically drive less than 6,000 km a year, are attracted to car sharing because, beyond this threshold, owning a car becomes more cost-efficient than sharing it. Thus one of the biggest motivations to use car sharing is saving money while retaining most of the flexibility offered by car (Kawgan-Kagan, 2015; Lane, 2005). This will apparently increase access to a more comfortable mode of transport for people with limited financial means or no car ownership. Kim (2015) notes that, even though car sharing is still considered an exclusive programme to middle-income, white, and young populations (Viechnicki *et al.*, 2015), the results of his study in New York City reveal that car sharing in low-income neighbourhoods did not differ from the typical car-sharing locations because the service was affordable. In the study *Shared Mobility and the Transformation of Public Transit* (American Public Transportation Association or APTA, 2016), the authors also confirm the same through noting that car sharing was the top alternative mode for low-moderate income residents and its use decreased with increasing incomes, but, interestingly, the use of bike sharing was preferred.

Shaheen *et al.* (2014) studied 23 bike-sharing programmes in North America and found that 43% of these programmes considered low-income groups in locating their stations. Other programmes also said that they aimed to serve the low-income groups currently or in the near future. The main obstacles identified for low-income groups were the need for debit/credit cards, minimum bank balance or deposit to cover for vandalism or theft. Bike-sharing companies are removing these prerequisites by working with funding agencies and local bodies to offer low-income population alternative ways to use these services such as prepaid cards, PayPal, cash, or through their telephone numbers. They even conduct outreach programmes to create awareness about sharing cycles, riding cycles, and so on.

Typically, the smart mobility services are web-based apps available on smartphones, which perpetuate the digital divide between those who use smartphones and those who don't. When we talk about smart solutions and services, we also need to discuss access to these solutions. Although there has been a considerable increase in smartphone users in most developing countries, and this continues to grow, the actual penetration level by population remains less than 50%. A bigger issue than access to digital services or a smartphone is the lack of digital literacy, the knowledge, comfort, and confidence to use smartphones. In many emerging economies, disparities in digital literacy compound disparities in basic literacy and reduce people's access to smart solutions and services. Governments need to tackle this digital literacy gap. For example, India launched the National Digital Literacy Mission (NDLM) in 2017 to make at least one adult from each household digitally literate by being able to utilise various kinds of technology, i.e. the internet, smartphones, computers, tablets, etc. Such steps can assist in reaping the benefits of smart mobility services and consequently create higher equity.

The role of governments in tackling the digital literacy gap and creating alternative methods of payment for low-income households is just one of the ways in which they can ensure equity of smart mobility services. As these services are predominantly offered by private players, the need for these services to be financially profitable is paramount, which also implies that they should mostly be available in locations that are dense or have high potential travel demand. Hence, spatial locations of these services might lead to inequitable access to people located in less dense neighbourhoods.

Another type of social segregation in selective approval on the grounds of race has also been observed in some cities. Thus, it comes down to the state to guarantee an appropriate degree of equity and non-discriminatory access to smart mobility services (Docherty *et al.*, 2017). As pointed out by various research studies (APTA, 2016; Docherty *et al.*, 2017; Lauwers and Papa, 2015; Viechnicki *et al.*, 2015), public bodies must work together with the smart mobility service providers in order to ensure that collective mobility goals are not treated as secondary to the individual/personal mobility goals.

SAFETY

Discussions around smart solutions for surveillance often involve CCTVs (Closed Circuit Televisions), which are presented as a default solution to crime prevention and reduction – a way for authorities to maintain 'eyes on the street' (Jacobs, 1961) without expending too many resources. It does appear to be a cost-effective way to maintain law and order in cities while reducing police presence. However, CCTV usage is heavily criticised as an instrument to privatise public space and strengthen the surveillance state. Critics argue that livelier streets and a porous public realm that promotes social cohesion are essential to safer cities, rather than digital surveillance and more security. Although some argue that CCTVs instil feelings of safety among citizens, a growing body of evidence reflects minimal impact of CCTV on perceptions of safety (Klauser, 2007; van Aalst *et al.*, 2014). Surveys reveal that, unless the CCTV footage is monitored live, which is not possible for all areas of the city 24 hours a day, CCTV cannot be effective in crime prevention. Its use will mainly be restricted to providing evidence in solving crimes, long after they have been committed.

Ride-sourcing (also known as 'on-demand-rides' or 'ride-hailing') is one of the most popular forms of smart mobility. However, there have been some considerable safety setbacks because cases of drivers sexually assaulting female passengers have emerged from across the globe. In response to such incidents, companies have started conducting background checks on drivers and, in some case, provide requisite training. This issue is, however, trickier than it seems. Under the model of the ride-sourcing services, drivers are often independent contractors who can be provided with incentives but cannot be subject to employment conditions (APTA, 2016). It must be noted that these efforts should have come as preventive measures rather than as a response to women's safety needs. All collective and public transport solutions must have elements of security inbuilt from the inception stage, rather than as an afterthought in the wake of violence against women. Given the emerging demand for safe transport services for women, women-only ride-hailing services (exclusively women drivers and passengers) have been launched in many countries, such as Riding Pink in Malaysia, LadyDriver, and FemiTaxi in Brazil, See-Jane-Go in the USA, and almost a dozen similar services in India. In Indonesia, where people often hitch rides, two women-only, motorcycle ride-hailing services, LadyJek and Sister Jek, were launched.

Some of the other safety checks that can suffer due to contractual relationships between drivers and ride-hailing companies are: alcohol and drug testing, liability and occupational safety training, higher vehicle safety, inspection, and insurance requirements. The smart mobility agenda does not currently cover these topics.

In view of the gaps in the smart mobility measures, it is clear that these systems need to be further refined to 'anticipate' and 'avoid' unsafe situations rather than 'responding' to the incidents and making amends at a later stage.

HEALTH

Digitalisation of the transport system may result in significant improvements in the realm of health. The application of smart technologies used in car sharing and allied services is built on an ethos of compact cities, where the need for a car in daily life will be substantially reduced. These compact cities encourage walking, cycling, and public transport usage, and entail both high usage of active modes and substantially reduced emissions.

A 2008 research study (Martin *et al.*, 2010) documented that, across a sample of approximately 9,500 participants spread over 7 cities in North America, 25% reported having sold a vehicle and 25% postponed a vehicle purchase due to car sharing. Shaheen *et al.* (2016) summarise about a dozen studies (conducted between 1986 and 2010) in the USA and Canada on car sharing, and found that it has led to a reduction in vehicular ownership in almost all cities. Some studies estimate that the introduction of one shared car, may lead to a reduction of four to over twenty owned cars (Martin *et al.*, 2010; Shaheen and Cohen, 2007; Sioui *et al.*, 2013). Giesel and Nobis (2016) studied the users of station-based and free-floating car-sharing services in two German cities (Berlin and Munich) and found that both types have respectively led to about 7% and 15% car shedding.

The car-sharing services also have an impact on the number of cars on the roads because more people use the same car in a sharing mode. Martin and Shaheen (2016) studied the impacts of a car-sharing service run by Car2go in 5 cities in North America, covering close to 10,000 Car2go members, and confirmed that, for every Car2go vehicle, the service creates a net reduction of about 11 cars on the roads. Similarly, in Philadelphia, each PhillyCarShare vehicle replaced an average of 23 private vehicles (Lane, 2005). This further reduces the number of vehicle miles travelled and greenhouse gas emissions. As the fixed costs associated with vehicle ownership are replaced by variable costs, people tend to become more conscious when deciding whether to make the trip at all and compare alternative transport modes (Cervero *et al.*, 2007; Duncan, 2011; Sioui *et al.*, 2013).

Car sharing has led to reductions in distances travelled by car (Cervero *et al.*, 2007; Martin *et al.*, 2010) – sometimes up to 44% reduction in vehicle miles or kilometres travelled (VMT/VKT) on average per user (Shaheen *et al.*, 2009) and CO_2 emissions in the range of 142–312 kg per person per year (Haefeli *et al.*, 2006; Martin and Shaheen, 2011). Another study in the USA on the impacts of car sharing (Chen and Kockelman, 2016) reveals that car-sharing members reduced their average individual transportation energy use and greenhouse gas emissions by approximately 51% due to modal shift, avoided travel, savings in parking demand, and fuel consumption. Similarly, in the Netherlands, a survey of car-sharing members found that 30% of the respondents owned fewer cars and drove 15–20% fewer kilometres (Nijland and van Meerkerk, 2017), resulting in between 240 and 390 kg of CO_2 fewer per person per year.

Considering the impact of smart mobility services on active mobility, studies have found that people who use public transit and other shared modes drive less and walk more (APTA, 2016; Kent, 2014; Lane, 2005). A study conducted in eight North American cities (Shaheen. *et al.*, 2014) found that regular users of bike sharing substituted other modes of transport with bike. Almost all cities reported a higher shift from public transport to cycles. The users of PhillyCarShare also reported driving less, walking and biking more, and using the transit more. Thus, solutions such as car sharing will contribute in a twofold manner: towards (1) a transition to low-carbon urban mobility and associated reduction in pollution levels, and (2) an increase in active mode usage leading to health benefits.

Sustainability of community

COHESION

The evidence given in earlier sections relating to reduction in car ownership, kilometres driven, greenhouse gas emissions, and increase in active mobility are indicators that smart mobility services have the potential to deliver sustainable results. The users of shared mobility have also shown change in their attitudes towards transportation. For example, PhillyCarShare members made travel decisions more judiciously, expressed greater awareness of transportation costs, and valued environmental friendliness far more than previously (Lane, 2005). A study on the attitudes of users of car sharing and their reasons for joining this service in North America reveals that people who are attracted to car sharing can be considered to be social activists, environmental protectors, innovators, economisers, or practical travellers (Burkhardt and Millard-Ball, 2006). Studies also mention that some users had a sense of belonging by being a part of a sharing community (Duncan, 2011; Litman, 2000; Roland Berger, 2014).

The APTA (2016) examined the issues and opportunities for transit systems in the wake of smart mobility, and found that people who use shared modes more frequently were more likely to use public transport, own fewer cars, and be judicious in their spending on transportation overall. The shared modes were found to substitute more automobile trips with transit trips. For better understanding of 'cohesion', there should be more in-depth life cycle assessment (LCA) of the benefits and how equally these benefits and costs are distributed over the society.

This outcome measure also includes 'aesthetics' and, although there is not much on this issue, there are reports on how dockless bike-sharing schemes can actually create inconvenience to non-users, and become a victim of rampant vandalism and an eyesore. What was meant to make it convenient for riders has become a source of great inconvenience for others. In many Chinese cities, for example Beijing, riders of dockless bike-share systems often leave their cycles on footpaths, residential colonies, and even in the middle of the road (Hernandez, 2017). Disgruntled and inconvenienced people have resorted to vandalising cycles by breaking their locks, hanging them on trees, throwing them in rivers, and even setting them on fire. As a response to this behaviour, companies have imposed fines for such behaviour and authorities have had to compound improperly parked cycles.

AWARENESS *(AVAILABILITY OF INFORMATION AND INFORMATIVE SIGNAGE)*

Smart mobility services are run by private companies and, in order to make a profit, they ensure that there is adequate information and outreach to inform people about these services. However, depending on their targeted customer group, based on income groups, race, neighbourhoods, etc., their efforts run the risk of being directed towards selected groups.

Furthermore, smart mobility solutions are dependent on smart phones which require digital literacy and, as mentioned earlier, access to services or availability of information is also hindered by the lack of digital literacy. Although most of the digitally illiterate people may also not have financial resources to access these services, elderly people often fall into this group of people who may be able to afford these services, need them, but have not kept up with the digital trends and technology.

Clearly, digitalisation comes with its own set of restrictions and cannot offer a singlehanded solution to the entire gamut of social sustainability challenges. However, if digital platforms are supplemented with formal follow-up procedures, and these procedures are based on both process and outcome measures, much can be achieved in the future. Though smart innovations in digital space holds great potential to make cities safer and equitable, they must include corresponding legal and policy changes, tangible changes in the urban built environment, and interventions targeting behavioural changes.

How can smart mobility be made socially sustainable?

Kitchin (2016) argues that smart city thinking and initiatives need to be reframed, reimagined and remade. His arguments are based on enabling better design and more equitable and just smart city initiatives. And, in these arguments, we see overlaps with the discussions presented on the social sustainability dimensions of the transport sector. Following the structure proposed by Kitchin (2016), we present six ways of reframing, reimagining, and remaking smart mobility.

The first covers goals that need to include normative questions about 'for whom and what purpose are smart mobilities being developed?', setting a wider framework within which smart mobility initiatives are formulated, developed, and run.

The second concerns conceptualising and contextualising smart mobility initiatives within a broader and richer understanding of the sort of transition we are looking for in the transport sector. This includes recognising that technical solutions will not work alone and must be positioned alongside and integrated with more social-, political-, legal-, and community-oriented solutions.

The third is through adding epistemological questions to big analytics, to acknowledge the situatedness, positionality, contingencies, assumptions, and shortcoming of data, and complement them with knowledge, understanding and explanations of how the different groups in a city move, hence recognising the multiple, complex, interdependent nature of a city.

These three ways concern normative and conceptual thinking of smart cities and smart mobility. In the following, we consider more practical and political ideologies and praxes.

The fourth way covers management and governance, implying that smart technologies that enable more efficient and effective transport need to be reframed into a blended, open, and co-produced form of urban management and governance.

The fifth concerns ethics and security. Privacy and security issues need to be considered and a balanced, pragmatic approach needs to be evolved, enabling the implementation of smart mobility solutions in a way that is gender and age sensitive.

The last way concerns stakeholders and working relationships. Transport is a contested subject, and often lies under the jurisdiction of several ministries and departments. The additional layer of 'smartness' further complicates this issue. Therefore, it is vital that smart mobility protagonists and various stakeholders need to be brought into the dialogue, with the aim of learning through co-creation and co-production.

It is essential that the smart mobility agenda disaggregates working tools on these guiding principles, and works towards bringing research and practice together.

From a methodological perspective, cities can initiate continuous mapping and analyses of urban mobility patterns and travel behaviour, with a special focus on different population groups. This task can utilise data collected from smartphones to analyse the travel patterns and further build a knowledge base of who can utilise new modes, such as new forms of shared mobility, Mobility-as-a-service (MaaS), ride-sourcing, etc. The digital or smart way of collecting travel data has opened new possibilities for analysing the conditions for promoting new forms of shared mobility, car sharing, MaaS, etc.

Travel data collected through smartphones can also assist in creating macro-based prognosis models for assessing the impact of travel trends and demographic changes on travel behaviour within regions. Such models can create a prognosis using current travel data, elasticities reflecting the interrelationship of modes, growth modelling, and the expected demographic development within specified geographical areas. Through combining the different elements, such models can give potential estimates of how trends in mode usage within certain age groups will affect the usage of competing modes and the overall modal distribution in an area. Methods such as these could become routine for municipalities and regional authorities based on regular feeds from different sources of big data and statistical bureaus.

Conclusion

This chapter has taken a look at the ethos of social sustainability in the transport sector and, furthermore, at the widespread movement towards the adoption of smart technologies in this sector. Smart mobility has been critiqued by applying a social sustainability filter to explore the exclusive nature of mobility outcomes. By doing so, we question the efficacy of smart transformations, and highlight the need for stronger governance to manage the adoption of smart technologies.

Through the case of ICT-led solutions, we have tried to highlight that the smart city agenda fails to recognise the exclusivity of solutions and the politics that surrounds mobility technologies. Although issues around smart mobility are starting to emerge alongside broader debates on the social impact of these technologies, the time is ripe for mainstreaming social sustainability in the planning and design of smart mobility. There is no doubt that smart technologies will provide the modus operandi of running the transport sector in the coming decades. It needs to be continuously emphasised that the mere adoption of technology and provision of smart services do not and will not necessarily lead to societal benefits. The social differences can further widen if only certain demographic groups can afford or access smart services. Application of technology should succeed education and awareness and not vice versa. Smart cities can be truly smart only if they are proactive, rather than reactive.

The meaning and importance of social sustainability differ for different implementing agencies. Smart solutions, discussed so far, have been essentially private projects and not public projects. They look at urban dwellers as 'consumers' of their service and are designed using a consumer-centric approach, where the collective social mobility goals and public interests are not given priority (Lauwers and Papa, 2015). These smart solutions initially came up as a response to latent demand for travel that was not being met by the traditional transit or transportation systems and, with the success of these models in one city, they were replicated in large numbers and larger scales in different cities around the world. Social sustainability assessment for such projects is often not required as per the formal procedures of getting permits for operation. However, they can be assessed using outcome measures, as we have provided here.

Often, the proponents of smart mobility solutions present unfounded claims not just to fulfil the service demand, but also to create social benefits of higher accessibility or affordability. For example, Hannon *et al.* (2016) claim that autonomous vehicles will provide improved accessibility by allowing people such as elderly people, handicapped people, or people without a driving licence. Researchers are discussing whether driving licences would become a thing of the past. However, we overlook the fact that physical accessibility is not real without affordability, and autonomous vehicles might have little trickle-down effect (Jeekel, 2017). Furthermore, mobility solutions that encourage single-occupancy vehicles can add to a bigger social problem of road congestion, leading to stress, waste of time, anxiety, and related concerns. Autonomous cars are also being posited as a solution to road accidents because they will eliminate 'human error' from driving, but this remains a spurious claim. However, given that this technology is close to the implementation stage, a social assessment using 'process' and 'outcome' measures should be prioritised.

The most essential takeaway from this chapter is the fact that transport needs to be decoupled from a purely technocratic assessment and fused with evaluations on social sustainability. Often the term 'social' emerges as a fuzzy concept without any clear structure for evaluation and implementation. In order to avoid this fuzziness, one can adopt and build upon evaluation methods like the transport rating systems and indicator-based assessments, informed by both quantitative and qualitative assessments, to give a structure to the evaluation processes that should be conducted at the macro-, meso-, and micro-levels.

Elements of transport systems like affordability, availability, and safety have direct and voluminous impacts on well-being, which need to be factored in from the very inception of a transport project. The first steps in designing a transport project should entail detailed assessments of 'social equity' and 'sustainability of communities' parameters, and existing projects should also be put under scrutiny to comment on its current status from the perspective of social sustainability.

Note

1 The programme

> aims to examine the concept of a smart-city in different geographical contexts, develop and establish Smart-city indicators, establish partnership and cooperation for action, promote exchange of best practices and knowledge transfer on sustainable urban development, including 'smart cities' and scale up the concept of smart cities and create a multiplier effect.
>
> (United Smart Cities [USC], available from https://sustainabledevelopment.un.org/partnership/?p=10009)

References

Albino V, Berardi U, and Dangelico RM (2015) Smart cities: Definitions, dimensions, performance, and initiatives. *Journal of Urban Technology* 22: 3–21.

American Public Transportation Association (APTA) (2016) *Shared Mobility and the Transformation of Public Transit*. Washington, DC: APTA.

Benevolo C, Dameri RP, and D'Auria B (2016) *Smart Mobility in Smart-City. Action Taxonomy, ICT Intensity and Public Benefits*. Geneva: Springer International Publishing.

Burkhardt JE and Millard-Ball A (2006) Who is attracted to carsharing? *Transportation Research Record* 1: 98–105.

Cervero R, Golub A, and Nee B (2007) City carshare: Longer-term travel demand and car ownership impacts. *Transportation Research Record: Journal of the Transportation Research Board* 1992: 70–80.

Chen TD and Kockelman KM (2016) Carsharing's life-cycle impacts on energy use and greenhouse gas emissions. *Transportation Research Part D: Transport and Environment* 47: 276–284.

Curz R, Kim JL, and Cha HS (2012) Using a thematic framework to compare sustainability rating systems applicable to transportation projects. In: *Construction Challenges in a Flat World*, Proceedings of the 2012 Construction Research Congress, pp. 1991–2000.

Docherty I, Marsden G, and Anable J (2017) The governance of smart mobility. *Transportation Research Part A: Policy and Practice* 115: 114–125.

Dondero G, Rodgers K, and Hurley P (2013) Developing a comprehensive sustainable transportation analysis framework. Paper presented at Transportation Research Board 92nd Annual Meeting (No. 13-0348).

Duncan M (2011) The cost saving potential of carsharing in a US context. *Transportation* 38: 363–382.

Giesel F and Nobis C (2016) The impact of carsharing on car ownership in German Cities. *Transportation Research Procedia* 19: 215–224.

Haefeli U, Matti D, Schreyer C, *et al.* (2006) *Evaluation Car Sharing*. Geneva: Bundesamt für Energie BFE.

Hannon E, McKerracher C, Orlandi I, *et al.* (2016) *An Integrated Perspective on the Future of Mobility*. McKinsey & Co. and Bloomberg New Energy Finance. Available at: http://worldmobilityleadershipforum.com/wp-content/uploads/2016/11/An-integrated-perspective-on-the-future-of-mobility.pdf.

Hernandez JC (2017) Shared bikes bring Chinese a 'monster-revealing' mirror. *New York Times – Asia Pacific* A1.

Holland R (2014), Critical interventions into the corporate smart-city. *Cambridge Journal of Regions, Economy and Society* 8: 61–77.

Jacobs J (1961) *The Death and Life of Great American Cities*. New York: Random House.

Jarvis H, Pratt AC, and Cheng-Chong Wu P (2001). *The Secret Life of Cities: The Social Reproduction of Cities*. Edinburgh: Pearson Education.

Jeekel H (2017) Social sustainability and smart mobility: Exploring the relationship. *Transportation Research Procedia* 25: 4296–4310.

Kawgan-Kagan I (2015) Early adopters of carsharing with and without BEVs with respect to gender preferences. *European Transport Research Review* 7.

Kent JL (2014) Carsharing as active transport: What are the potential health benefits? *Journal of Transport & Health* 1: 54–62.

Kim K (2015) Can carsharing meet the mobility needs for the low-income neighborhoods? Lessons from carsharing usage patterns in New York City. *Transportation Research Part A: Policy and Practice* 77: 249–260.

Kitchin R (2016) Reframing, reimagining and remaking smart cities. The Programmable City Working Paper 20. Creating smart cities, workshop hosted by the Programmable City project at Maynooth University, Ireland, 5–6 September 2016. Available at: https://osf.io/cyjhg.

Klauser FR (2007) Difficulties in revitalizing public space by CCTV. *European Urban and Regional Studies* 14: 337–348.

Lane C (2005) PhillyCarShare: First-year social and mobility impacts of carsharing in Philadelphia, Pennsylvania. *Transportation Research Record: Journal of the Transportation Research Board* 1927: 158–166.

Lauwers D and Papa E (2015) Towards a smarter urban mobility. Colloquium Vervoersplanologisch Speurwerk, Antwerp, Belgium.

Lineburg K (2016) Transportation rating systems and social sustainability: A comprehensive analysis. Senior Honours Projects. Paper 167. James Madison University.

Litman T (2000) Evaluating carsharing benefits. *Transportation Research Record* 31–38.

Litman T (2012) *A New Social Equity Agenda for Sustainable Transportation*. Victoria, Australia: Transport Policy Institute. Available at: www.vtpi.org/equityagenda.pdf.

Lucas K, Marsden G, Kimble M, and Brooks M (2007) Assessment of capabilities for examining long-term social sustainability of transport and land use strategies. *Transportation Research Record* 2013: 30–37.

Lucas K and Stanley J (2013) Achieving socially sustainable transport in the developing context. Paper presented at the 13th World Conference on Transport Research, 15–18 July, Rio De Janeiro, Brazil.

Manaugh K, Madhav B, and El-Geneidy A (2015) Integrating social equity into urban transportation planning: A review of metropolitan transportation plans in North America. *Transport Policy* 37: 167–176.

Marsden G, Kelly C, Nellthorp J, Lucas K, and Brookes M (2007) *Appraisal of Sustainability in Transport: Final Report*. Leeds: Institute for Transport Studies, Leeds University. Available at: www.its.leeds.ac.uk/projects/sustainability/resources/Sustainability_Appraisal_Final Report.pdf.

Martin EW and Shaheen SA (2011) Greenhouse gas emission impacts of carsharing in North America. *IEEE Transactions on Intelligent Transportation Systems* 12: 1074–1086.

Martin EW and Shaheen S (2016) *Impacts of Car2go on Vehicle Ownership, Modal Shift, Vehicle Miles Traveled, and Greenhouse Gas Emissions: An Analysis of Five North American Cities*. Transportation Sustainability Research Center (TSRC).

Martin EW, Shaheen SA, and Lidicker J (2010) Impact of carsharing on household vehicle holdings: Results from North American shared-use vehicle survey. *Transportation Research Record*: 150–158.

Mercier J (2009). Equity, social justice, and sustainable urban transportation in the twenty-first century. *Administrative Theory & Praxis* 31(2): 145–163.

Newman P (2016) The '30-minute city': How do we put the political rhetoric into practice? *The Conversation*. Available at: http://theconversation.com/the-30-minute-city-how-do-we-put-the-political-rhetoric-into-practice-56136 (accessed 18 March 2018).

Nijland H and van Meerkerk J (2017) Mobility and environmental impacts of car sharing in the Netherlands. *Environmental Innovation and Societal Transitions* 23: 84–91.

Roland Berger (2014) Shared mobility: How new businesses are rewriting the rules of the private transportation game. *Think Act*. Munich: Roland Berger Gmbh.

Shaheen S and Cohen A (2007) Growth in worldwide carsharing: An international comparison. *Transportation Research Record*: 81–89.

Shaheen S, Cohen A, and Zohdy I (2016) *Shared Mobility: Current Practices and Guiding Principles*. US Department of Transportation, Federal Highway Administration.

Shaheen SA, Cohen AP, and Chung MS (2009) North American carsharing: 10-year retrospective. *Transportation Research Record* 35–44.

Shaheen SA, Martin EW, Chan ND, *et al*. (2014) *Public Bikesharing in North America During a Period of Rapid Expansion: Understanding Business Models, Industry Trends and User Impacts*. San Jose, CA: Mineta Transportation Institute, p. 234.

Sioui L, Morency C, and Trépanier M (2013) How carsharing affects the travel behavior of households: A case study of Montréal, Canada. *International Journal of Sustainable Transportation* 7: 52–69.

van Aalst I, Schwanen T, and van Liempt I (2014) Video-surveillance and the production of space in urban nightlife districts. In: van den Hoven J, Doorn N, Swierstra T, *et al*. (eds), *Responsible Innovation 1: Innovative Solutions for Global Issues*. Dordrecht: Springer Netherlands, pp. 315–334.

Viechnicki P, Khuperkar A, Fishman TD, *et al*. (2015) *Smart Mobility: Reducing Congestion and Fostering Faster, Greener, and Cheaper Transportation Options*. New York City: Deloitte Touche Tohmatsu Ltd, p. 38.

Whitzman C (2017) A 20-minute city sounds good, but becoming one is a huge challenge. *The Conversation*. Available at: https://theconversation.com/a-20-minute-city-sounds-good-but-becoming-one-is-a-huge-challenge-80082 (accessed on 18 March 2018).

5 Social sustainability and urban heritage

The challenge of conserving physical places and sustaining cultural traces

Chris Landorf

Introduction

Although cities are increasingly required to respond to the challenges of sustainable development, there is a sufficient body of literature to suggest that the strategies available to address these challenges remain vague and complex to operationalise (Dalmas *et al.*, 2015; Klopp and Petretta, 2017; van Oers and Periera Roders, 2017). This is particularly so in relation to the social dimension of sustainability, with its reliance on hybrid sets of hard-to-measure 'soft' indicators and emphasis on ambiguous concepts such as governance, community and culture (Colantonio and Dixon, 2011; Shirazi and Keivani, 2017). As this chapter argues, the challenge becomes even more complex when applying strategies for social sustainability to historic urban cultural landscapes.

For sensationalists, the ascendance of urban dwelling has been a magnet for predictions that cities will become Blade Runner-esque dystopias where civil society hangs by a thread and, the 'neoliberal project' has been naturalised 'on a global scale' (Jessop, 2002: 452). In these cities of the future, market capitalism and the cult of the individual will prevail over social democracy and communities that identify with a common heritage. For pragmatists, the urbanisation of the world's population is less the consequence of post-apocalyptic necessity and more the result of cities being the engines of wealth generation, technical innovation, and cultural creativity. In these cities of the future, social sustainability will be underpinned by a participatory approach to urban heritage that reflects pluralistic social values and accepts a continual process of cultural construction and reconstruction.

With the latter of these scenarios in mind, this chapter aims to examine the specific challenges facing historic urban cultural landscapes where balancing sustainable development must also consider conserving heritage significance. The argument to be advanced is initially framed in an understanding of what heritage is and the role that it plays in the vitality of cities. The discussion is then grounded in an appreciation of the contradictions that sustainable development and, more specifically, social sustainability presents for historic urban cultural landscapes. With this understanding, the chapter then explores recent efforts to transform conventional heritage principles into tools for sustainable development. The chapter concludes with an analysis of the extent to which the principles of sustainable development have been integrated into management practices at four World Heritage urban landscapes.

The historic urban cultural landscape

It is important at this point to introduce some definitions that will be used to guide the discussion to follow, starting with the difference between 'history' and 'heritage'. History and the historic, Halbwachs ([1926] 1992) argues, lay claim to an objective and universalising authority that distinguishes between past and present. Heritage, on the other hand, is associated with what Smith (2006: 59) describes as the 'subjective and unselfconscious workings' of individual and collective memory. Unlike professional histories, memory is integral to identity formation and it consequently has an intensely emotive power. This is particularly so when collective memory is associated with nationalism and the official practices used to construct national identity and legitimise cultural dominance.

Also exploring the interrelationship of heritage, memory, and identity, Graham *et al.* (2000: 34) describe the appropriation, privileging, and invention of heritage as 'intimately related to the exercise of power'. Crucially, heritage is linked to practices of social inclusion and, by association, exclusion (Ashworth and Tunbridge, 1996; Harrison, 2010; Hobsbawm and Ranger, 1983; Lowenthal, 1998). Whether consciously manufactured or unconsciously formed, these practices are enacted by way of the relationship that heritage has with the present through the act of remembering. Smith (2006) describes remembering as a dynamic cultural process in which the past is continually negotiated and reinterpreted based on the needs of the present. She also draws attention to the connection between the 'representational and symbolic value of heritage in constructing and giving material reality to identity' (Smith, 2006: 48). The association between memory and identity, and heritage as a physical representation of these concepts, is exemplified in the modern process of listing places and practices considered to be of heritage significance (Harrison, 2013). The contradictory implications of this for socially sustainable urban heritage are returned to later.

A further differentiation in keeping with the focus of this chapter is between the terms 'historic environment' and 'historic cultural landscape'. As Gibson (2009: 71) reasons, the term 'historic environment' conveys a static understanding of a physical canvas 'on which the evidences of history can be discerned', whereas 'historic landscape' suggests a dynamic cultural setting where meanings are constantly being constructed and re-constructed. Taylor (2015: 184) also draws a distinction between the image of an historic urban environment as 'a closed view of architectural wonders', whereas an historic cultural landscape is an experience, reflective of both intangible attributes and spatial phenomena deposited by diverse communities over time. Historic cultural landscapes suggest layers of the physical components and patterns that distinguish cities, as well as the symbolic meanings, sociocultural values, behaviours, and traditions that give rise to a collective local identity. The concept of a dynamic cultural setting is stressed here because it resonates with the social sustainability principles of equity, participation, inclusion, diversity, and sense of place.

The focus of this chapter is also on historic 'urban' cultural landscapes. The term is used here to describe the physical and intangible characteristics that distinguish those historic areas of a city that also contain higher density, socioeconomically diverse human populations. As Dovey (2016) contends, urban space plays a key role in concentrating activity and intensifying encounters with difference, both central to achieving economic vitality, social inclusion, and diversity. He suggests that an assemblage of density, mix, and access are the necessary core of urban intensity, and that these dimensions are spatially framed. The paradox lies in there being a strong tradition in understanding and valuing the formal composition of buildings and public spaces, but not the informal design, use, and meanings associated with urban space. Furthermore, a city is 'created by many hands rather than a single controlling vision' and perceived as an experience 'rather than submitting to our gaze like a beautiful object or scene' (Dovey, 2016: 14). This aspect of the city as a generator of vitality is discussed further below in relation to social sustainability, diversity, and equity.

This thinking reflects growing international recognition that urban heritage can no longer be managed in isolation from the prevailing forces acting elsewhere on a city. The 'historic city is not an island', Bandarin (2015: 11) observes, the 'barriers created by special legislation and programmes aimed at its protection are unable to shield it … from the rest of the city'. Therefore, although cities with centres recognised for their cultural significance may have advantages in terms of their inherent heritage value, the increasingly accepted view is that the economic, environmental, and social dimensions of historic urban cultural landscapes need to be re-conceived as integral parts of larger wholes. The city as an evolving cultural setting is also discussed further below in relation to social sustainability and authenticity.

The contradictions of social sustainability for urban heritage

This evolving discussion suggests that, to be socially sustainable, historic urban cultural landscapes require definitions of significance that are culturally inclusive yet specific enough to capture those tangible and intangible characteristics that are of heritage value. They require governance models that embrace a holistic and strategic orientation, and ongoing participation in the definition and management of heritage significance. Finally, they require the traditional focus on the conservation of physical fabric to be broadened to include cultural practices, and to respect change as an inherent feature of those practices.

This represents a departure from established planning and conservation traditions where iconic buildings and public spaces were preserved 'in an isolated manner, divorced from their wider urban context', including cultural practices (Labadi and Logan, 2016: 2).

The chapter has argued thus far that cultural heritage is as intrinsic to sustainable urban development as combating poverty and changing consumption patterns. The United Nations suggests, however, that urban sustainability will require the generation of better employment opportunities, equitable access to basic infrastructure, a reduction in the number of people living in slums, and preservation of the 'natural assets within cities and surrounding areas' (United Nations, 2014: 17). Critically, this summary omits the cultural assets of a city – that unique combination of static canvas and dynamic setting that accords the opportunity for economic vitality, social diversity, and the production of cultural meaning (Gibson, 2009). As the following discussion indicates, in addition to the recognition of cultural assets as an integral dimension of sustainable urban development, there are two contradictions that need to be managed in historic urban cultural landscapes. Both relate to the unique spatial qualities and functional densities of historic cities that also attract tourism.

The first contradiction concerns the impact of tourism on social–economic diversity and equity in historic urban cultural landscapes. At the extreme end, the tangible heritage places and intangible cultural practices that shape the attraction of historic cities for tourists are converted into commercial opportunities designed to target these transient populations. At the same time, property values rise in response to increased investment and the evolving attractiveness of inner-city heritage character neighbourhoods (Licciardi and Amirtahmasebi, 2012). As Bandarin (2015) points out, not all historic cities are affected in this way and not all gentrification is the result of market forces alone. For those interested in social sustainability and urban heritage, however, increased tourism activity invariably gives rise to the displacement of lower-income residents, a reduction in social–economic diversity, and an increase in city-wide inequity in terms of access to services and participation in civic society.

The second contradiction concerns the impact of tourism on historic urban landscapes as settings for the construction and reconstruction of cultural meaning and identity. Authenticity is discussed in greater detail shortly, but the concept is central to tourist expectations. The crux of the issue, as Urry (1995) persuasively argues, is that where 'authenticity' becomes something necessary to achieve economic success, local communities will adapt, reconstruct or 'freeze' heritage to meet tourist expectations. The commodified heritage product, object, or practice is adjusted to tourist demands for education, entertainment, and a safe experience (Chhabra *et al.*, 2003; Darlow *et al.*, 2012; Malcolm-Davies, 2004). The contradiction in this tourism–heritage dynamic is that it encourages the previously discussed preservation of 'authentic' products, objects, and practices in isolation from the rest of the city;

physically suspended at a particular point in time and culturally detached from changing traditions found elsewhere in the wider urban context.

Integrating social sustainability into urban heritage management

The discussion so far has framed an understanding of what historic urban cultural landscapes are and the challenges they present for social sustainability. The chapter now explores how social sustainability has been discussed within the heritage literature before moving to examine efforts to transform the international principles that guide the management of heritage into a more effective framework for sustainable development. Substantial attention has been paid to exploring the contribution that urban heritage makes to sustainable development (Bandarin and van Oers, 2012, 2015; Labadi and Logan, 2016; van Oers and Pereira Roders, 2014) and arguing the policy limitations of traditional economic techniques for the valuation of heritage (Dalmas *et al.*, 2015; Gibson and Pendlebury, 2009; Throsby, 2016; Wright and Eppink, 2016). This literature establishes the importance of urban heritage for the advancement of sustainable development and the difficulty of valuing heritage, particularly the intangible characteristics of heritage.

Further attention has been given to identifying sustainable development strategies and indicators for historic landscapes (Guzman *et al.*, 2017; Klopp and Petretta, 2017; Rey-Perez and Avila, 2017; Sowinska-Swierkosz, 2017). This literature recognises that heritage is an important, but difficult to quantify, medium for sustainable development policy. Additional consideration has been given to the principles of social sustainability as they relate to the built environment (Shirazi and Keivani, 2017) and, more specifically, to historic landscapes (Colantonio and Dixon, 2011; Landorf, 2009a, 2011; Yung and Chan, 2012). Argued throughout this literature is the strong link between historic landscapes and social sustainability, although strategies that encourage participation, inclusion, diversity, and a sense of place are also supported as a means to enhance equity and quality of life through heritage. What is noted as lacking, however, are practical implementation strategies, systematic assessment methodologies, and universal benchmarks. Also problematic is the absence of a central judicial body with the capacity to enforce international agreements on sustainable development in historic urban cultural landscapes, an issue that is returned to later.

This body of academic literature is reflected in the evolution of the conservation principles and guidelines adopted by key international organisations, such as the United Nations Educational, Scientific and Cultural Organisation (UNESCO) and the International Council on Monuments and Sites (ICOMOS). The landmark *Convention Concerning the Protection of the World Cultural and Natural Heritage* (the Convention) was adopted by UNESCO in 1972 with the primary aim of protecting places considered to be under threat of destruction. The concept of intergenerational equity, at the centre of sustainable development, is also evident in the Convention with Article 4 requiring each signatory nation, or State Party, to identify, protect, conserve, present, and transmit heritage to future generations. The Convention is noteworthy for establishing an intergovernmental World Heritage Committee and a World Heritage List of properties considered to be of 'outstanding universal value'. The ten criteria that make a property worthy of inscription were articulated in the first edition of the *Operational Guidelines for the Implementation of the World Heritage Convention* (the Operational Guidelines) in 1977. Although the Convention is a static legal instrument, objectives and procedures are occasionally revised through the Operational Guidelines.

A nominated property must meet at least one criterion for outstanding universal value, as well the conditions of authenticity and/or integrity, and have an adequate system of protection and management (UNESCO, 2015). Inscription is based on a property's outstanding universal value, determined through comparison with similar properties around the world. Consideration is given, first, to a property's outstanding universal value, second, to the state of a property's authenticity and/or integrity in relation to the criteria under which it has been nominated, and, finally, to whether a property is being and will continue to be adequately protected. This final point includes not only having appropriate legislative frameworks to effect protection, but also access to technical and managerial expertise.

Outstanding universal value and authenticity are both imprecise concepts that present challenges for the evaluation and management of heritage properties (Cleere, 1996; Forrest, 2010; Smith, 2006; Stovel, 1994). Although the Convention is silent on the issue, the internationally influential Australia ICOMOS *Charter for Places of Cultural Significance* (the Burra Charter) provides some clarity by defining the cultural significance of a place as the 'aesthetic, historic, scientific, social or spiritual value for past, present and future generations'. Furthermore, there may be 'a range of values for different individuals or groups' and 'significance may change over time and with use' (Australia ICOMOS, 2013: 2). Throsby (2010) provides additional distinction between the economic and cultural values of heritage. Economic value, he argues, can be measured using recognised valuation techniques. Cultural value, however, 'has no single unit of account that can capture its multidimensional nature' (Throsby, 2010: 111). Heritage significance can, therefore, be viewed as intrinsically embodied in a place, object, or practice, giving it a cultural worth that is beyond valuation. Heritage value, on the other hand, is a description of measurable use and non-use values.

The 2002 *Budapest Declaration* was the first official UNESCO document to link sustainable development and heritage. The Budapest Declaration places particular emphasis on strengthening local participation and the flow-on of economic benefits by ensuring:

> an appropriate and equitable balance between conservation, sustainability and development, so that World Heritage properties can be protected

> through appropriate activities contributing to the social and economic development and the quality of life of our communities.
>
> (UNESCO, 2002: Article 3)

The Declaration also expresses the need to:

> ensure the active involvement of our local communities at all levels in the identification, protection and management of our World Heritage properties.
>
> (UNESCO, 2002: Article 3)

The 2005 *Vienna Memorandum* was adopted to promote the integration of 'contemporary architecture, sustainable urban development and landscape integrity based on existing historic patterns, building stock and context' (UNESCO, 2005: Preamble). The Memorandum acknowledges, for the first time, that change is an inherent tradition of cities.

In 2011, UNESCO adopted the *Recommendation on the Historic Urban Landscape* following a period of extensive consultation and in response to growing concern over the threats posed by development (Labadi and Logan 2016). The Recommendation:

> provides the basis for a comprehensive and integrated approach for the identification, assessment, conservation and management of historic urban landscapes within an overall sustainable development framework.
>
> (UNESCO, 2011: Article 10)

The Recommendation introduces the term 'historic urban landscape' and extends the historic city centre concept 'to include the broader urban context and its geographical setting' (UNESCO, 2011: Article 8). The Recommendation also identifies the need to develop a range of tools to support the management of sustainable historic urban landscapes. The tools, dealing with civic engagement, knowledge and planning, regulatory systems, and finance, reflect the need for local participation, the identification of heritage values that reflect diversity, and the 'management of change to improve quality of life' (UNESCO, 2011: Article 24). The tools have since been developed and tested in place-specific case studies (Bandarin and van Oers, 2012, 2015). The 2013 *Hangzhou Declaration* reaffirmed 'that culture should be considered to be a fundamental enabler of sustainability' and that 'a vibrant cultural life and the quality of urban historic environments are key for achieving sustainable cities' (UNESCO, 2013: 9).

Two final links are worth noting in this evolving chain of declarations and recommendations concerning sustainable development and urban heritage. In 2014, ICOMOS adopted the *Florence Declaration on Heritage and Landscape as Human Values*. The Florence Declaration recommends that community-driven conservation and local empowerment play an essential role in sustainable local socio-economic development (ICOMOS, 2014: Article 4). The World Heritage Convention and Operational Guidelines confirm a central role for ICOMOS in the provision of expert advice, as well as the evaluation of properties nominated for inscription and the ongoing monitoring process. As an international organisation for heritage professionals, ICOMOS also has significant influence at national and local levels.

In 2015, the UN adopted *Transforming Our World: The 2030 Agenda for Sustainable Development*. The Agenda created, for the first time, a direct relationship between cultural heritage and the sustainability of urban environments. Of 17 Sustainable Development Goals, Goal 11 aims to 'make cities and human settlements inclusive, safe, resilient and sustainable' with Target 11.4 being to 'strengthen efforts to protect and safeguard the world's cultural and natural heritage' (UN, 2015: 14 and 22).

This activity reflects efforts on the part of the international community to advocate for more effective links between sustainable development and cultural heritage. It also demonstrates the fundamental role assumed in that effort by urban heritage, and the focus placed on the social dimension of sustainability, primarily through strategies that encourage participation, inclusion, diversity, and sense of place as a means to enhance equity and quality of life. However, attempts to integrate the principles of sustainable development are complicated by the geo-politicisation of heritage, the self-interest of independent nation states, and the lack of a central authority with the power to legally enforce the Convention. As Harrison (2010: 191) argues, 'once heritage moves into the political arena it becomes a symbol of … nationalism, culture and class'. The semantic structure of the Convention and associated guidelines, together with localised heritage discourses, additionally work against operational progress.

Methodology

Evidence for the remainder of this chapter comes from a qualitative content analysis of two key documents from four World Heritage Listed properties identified as historic urban cultural landscapes: the nomination evaluation prepared by ICOMOS and the management plan prepared by the relevant State Party. Nominations for inscription include a description of the property, a justification for inscription, statements of outstanding universal value, integrity and authenticity, the framework for protection and management, a comparison against similar properties, and a summary of factors affecting the property (UNESCO, 2015: Annex 5). Nomination evaluations provide recommendations on inscription and comments against the key issues described in the nomination by the relevant State Party.

Management plans were adopted as the second source of evidence because of their role as the primary instruments guiding the 'protection, conservation, management and presentation' of World Heritage properties (UNESCO, 2015: Article 119). Regardless of the limited guidance available for their preparation, a management plan is a requirement of the nomination process (UNESCO, 2015: Article 108). Even so, many properties do not have one lodged with UNESCO. Furthermore, State of Conservation Reports prepared on properties under threat regularly acknowledge weak management systems and planning as a factor in poor outcomes. Many plans are additionally written for awareness raising, as much as for achieving World Heritage status (Rodwell, 2007). As such, they have a high international profile evinced by the numerous studies that have focused on heritage management processes (Blandford, 2006; Darlow *et al.*, 2012; Landorf, 2009a, 2009b; Poulios, 2014; Rodwell, 2002; Wilson and Boyle, 2006), and the development of indicators for monitoring purposes (Gullino *et al.*, 2015; Guzman *et al.*, 2017; Landorf, 2011; Sowinska-Swierkosz, 2017; Throsby, 2016).

As this research relied on text-based sources, and existing theory could be identified in relation to the dimensions of sustainable development and heritage management, a three-stage, directed-content analysis approach was adopted (Hsieh and Shannon, 2005). The first stage involved identification of those dimensions that should be evident in the documents used to guide the sustainable management of historic urban cultural landscapes. Two conceptual dimensions of sustainable development were consistently evident in the literature: the use of a long-term and holistic strategic planning process, and the participation and empowerment of multiple stakeholders in that process. As a management tool, strategic planning adopts a circular model of cause and effect to promote a holistic and future-oriented approach to resource management. As a management strategy, the meaningful engagement of multiple stakeholders throughout decision-making is pivotal to achieving a collective sense of ownership and responsibility. The literature also revealed consensus in terms of the component elements of these two conceptual dimensions. Following a framework developed by Landorf (2009a, 2009b, 2011) to assess the extent that the principles of sustainable development had been integrated into management practices at industrial World Heritage sites, the following four component elements or open coding dimensions were derived from the two conceptual dimensions of sustainable development: the situation, the strategic orientation, the stakeholder values, and the stakeholder participation dimensions.

The *situation analysis* dimension was used to determine whether external and internal influences on a heritage property were identified as a starting point in the planning process. This dimension responds to the need to establish a holistic understanding of the trends and issues impacting on a property, including social inclusion and diversity attributes, equity and infrastructure characteristics, and quality-of-life benchmarks. The *strategic orientation* dimension was used to establish whether the planning process was long term and holistic. A strategic framework responds to the dynamic nature of the influences on a property, bringing diffuse resources together effectively and providing stakeholder benefits equitably. The *stakeholder values* dimension was used to establish whether socially inclusive and diverse values, needs and expectations were integrated into a strategic vision for each property. This dimension resolves who the legitimate stakeholders are and is integral to enhancing a collective sense of identity and sense of place. The *stakeholder participation* dimension was used to determine the breadth and depth of stakeholder engagement in the planning and management process. Participation establishes a collective vision for the property and empowerment through this process is needed to effect sustainable change.

The second stage of the research involved the identification of an appropriate sample frame for analysis. World Heritage properties were selected based on the inscription process being subject to independent evaluation against internationally agreed criteria. The ongoing state of conservation is also subject to continuous monitoring. World Heritage properties are prestigious, they embody national identity, and have access to international funding and expertise. As a result, they are argued here to represent a benchmark of heritage management practice. To enhance the validity of the findings through cross-case comparison, the sample included a range of older and more recently inscribed urban landscapes from different continents. To further enhance cross-case comparison, the sample included varying levels of property maturity in terms of tourism infrastructure and market demand. The sample was further limited to properties that had current and publicly accessible management plans.

Case study

The World Heritage properties selected for analysis were the Historic Centre of Florence (Italy) inscribed in 1982 (Francini *et al.*, 2016; ICOMOS, 1982), Lamu Old Town (Kenya) inscribed in 2001 (Anonymous, n.d.; ICOMOS, 2001), Melaka and George Town, Historic Cities of the Straits of Malacca (Malaysia) inscribed in 2008 (Anonymous, 2008; ICOMOS, 2008), and the Historic City of Ahmadabad (India) inscribed in 2017 (Ahmadabad Municipal Corporation, 2016; ICOMOS, 2017). Rio de Janiero, inscribed in 2012, was the only World Heritage Listed cultural urban landscape in the Americas that had a management plan. It was not included, however, due to a significant area of natural landscape being included within the property boundary. Table 5.1 shows the criteria against which the four properties are inscribed on the World Heritage List. All four include intangible cultural practices and values, or criteria ii, iii, v or vi, as intrinsic to their World Heritage status.

The importance of intangible cultural heritage is further evident in the Statement of Significance for each property. For all four properties, significance is based on a combination of historical importance and the authenticity of surviving physical evidence. In addition, all properties include reference to some form of living sociocultural tradition. The significance of each property is summarised in Table 5.2.

The opening discussion argued the need to capture tangible and intangible values, embrace participation, and respect change as an inherent part of the

Table 5.1 Justification for inclusion of each property on the World Heritage List

World Heritage criteria (date of inscription)	*Florence (1982)*	*Lamu (2001)*	*Melaka (2008)*	*Ahmadabad (2017)*
i Masterpiece of human creative genius	✓			
ii Interchange of human values on developments in architecture or technology, monumental arts, town-planning or landscape design	✓	✓	✓	✓
iii Testimony to a cultural tradition or civilisation which is living or has disappeared	✓		✓	
iv Example of a type of building, architectural or technological ensemble or landscape	✓	✓	✓	
v Example of a traditional human settlement, land-use or sea-use				✓
vi Associated with events, traditions, ideas, beliefs, artistic and literary works	✓	✓		

Source: adapted from ICOMOS World Heritage evaluation for each property.

Table 5.2 Summary of the outstanding universal value of each property

Property	*Significance*
Florence	A unique social and urban complex that influenced the development of architecture and the arts in Europe through sustained creativity, a testament to its power as a Renaissance merchant city
Lamu	The oldest and best-preserved Swahili settlement in East Africa, an important trading port and religious centre that has maintained a unique physical and cultural integrity
Melaka and George Town	A unique example of multi-cultural trading port townscapes that bear testimony to the fusion of architectural and living cultural traditions of multi-culturalism in Asia
Ahmadabad	A unique walled city that exhibits an architectural fusion of Islamic culture and local traditions and crafts, a testament to the articulation of commercial functions and religious coexistence

Source: adapted from ICOMOS World Heritage evaluation for each property.

historic tradition of cities to achieve socially sustainable urban heritage. Yet, despite sociocultural values featuring in the significance of all properties, the frameworks for protection and management emphasised the conservation of physical components as the principal means of protecting World Heritage significance. Also relevant to the protection and management of each property is the complex overlay of national, provincial, and/or local government legislation and administering authorities. All properties fell under overarching national legislation that designates conservation status while management fell primarily to a local authority.

Discussion

Despite recognition of the importance of an integrated approach to planning and management, the key documents for all four properties failed to articulate clear organisational structures, management responsibilities, and links to national sustainable development frameworks. One property indicated that a dedicated World Heritage property coordinator was in place. The four management plans generally followed a conventional three-part structure: property description, evaluation of issues and objectives, and property management framework. None of the plans detailed how ongoing stakeholder collaboration was managed and limited information was provided in relation to community consultation during the plan preparation. An emphasis on heritage-led economic regeneration and tourism was evident in all four management plans, despite limited evidence to support demand.

The *situation analysis* included ten coding items. All four management plans provided historical backgrounds, inventories, and maps, and identified property-specific factors such as visitor management, transportation, and conservation as key management issues. Despite the lack of evidence, all assumed economic benefits would flow from increased tourism and only Florence mentioned generic visitor statistics and economic characteristics. None of the plans indicated that targeted research had been conducted into local socio-economic attributes and infrastructure characteristics. None considered the impact of broader national or international trends, and none detailed mechanisms for the regular evaluation of external factors as part of the planning process. Each property, therefore, remained relatively isolated from its context and focused on site-specific conservation issues.

The *strategic orientation* dimension included ten coding items. A linear planning process was evident in all plans with a vision, objectives, and action plans derived from an initial property evaluation and some form of stakeholder consultation. Although there was no evidence that a range of strategic alternatives had been evaluated, two plans did define objectives that could be described as broad based, i.e. not heritage focused. The need to develop adequate indicators to monitor progress was mentioned as a key management issue in all plans, but quantifiable measures for issues such as participation and social well-being were notable omissions.

The *stakeholder values* dimension included five assessment items. One plan identified local values, attitudes and issues, but none assessed quality of life or indicated that the vision for the World Heritage property aligned with community values. The *stakeholder participation* dimension included nine coding items. Despite recognition of the need for public participation, three plans were prepared as a consultation draft by a managing partnership group before wider public consultation was sought. One plan conducted stakeholder focus groups, but there was no evidence that this was extended to include the public at large, that it was an ongoing process, or that the issues raised were influential. Table 5.3 provides a summary of the evaluation results for each World Heritage property management plan.

The results indicate that the World Heritage nomination documents and management plans analysed in this study did not meet all the criteria for the integration of social sustainability principles into the heritage planning and management process. However, the two newer management plans for the Historic Centre of Florence and the Historic City of Ahmadabad both indicated that work was being done within the situation analysis dimension to establish existing land use and demographic characteristics. Florence and Ahmadabad also exhibited a strategic orientation that targeted the spread of heritage benefits more equitably within their respective communities. Florence was the only property where diverse values, needs, and expectations had been sought, but not necessarily integrated into the strategic vision and long-term planning for the property. As noted previously, this is integral to enhancing a collective sense of identity. Finally, all properties demonstrated

Table 5.3 Summary of the evaluation results for each property

Sustainability dimension and coding item		*Property (date of management plan)*			
Key: shaded = Evident/somewhat evident; unshaded = Not evident		*Florence (2016)*	*Lamu (2013)*	*Melaka (2008)*	*Ahmadabad (2016)*
Situation analysis	1 Tangible heritage described	Evident	Evident	Evident	Evident
	2 Intangible heritage described	Evident	Evident	Not evident	Evident
	3 Land use and ownership identified	Evident	Not evident	Not evident	Evident
	4 Demographic characteristics identified	Evident	Not evident	Not evident	Evident
	5 Economic characteristics identified	Evident	Not evident	Not evident	Not evident
	6 Economic benefits of heritage identified	Not evident	Not evident	Not evident	Not evident
	7 Heritage tourism activities identified	Evident	Evident	Evident	Evident
	8 Capacity of infrastructure identified	Not evident	Not evident	Not evident	Not evident
	9 Visitor details identified	Evident	Not evident	Not evident	Not evident
	10 Integration with other planning processes	Not evident	Not evident	Not evident	Not evident
Strategic orientation	11 Long-term orientation	Evident	Evident	Evident	Evident
	12 Broad-based economic goals	Evident	Not evident	Not evident	Evident
	13 Broad-based environmental goals	Evident	Not evident	Not evident	Evident
	14 Broad-based social/community goals	Evident	Not evident	Not evident	Evident
	15 Broad-based heritage goals	Evident	Evident	Evident	Evident
	16 Strategic alternatives considered	Not evident	Not evident	Not evident	Not evident
	17 Objectives support goals	Evident	Evident	Evident	Evident
	18 Objectives based on capability	Evident	Evident	Evident	Evident
	19 Objectives target equitable distribution	Evident	Not evident	Not evident	Evident
	20 Objectives are quantifiable	Not evident	Not evident	Not evident	Not evident

continued

Table 5.3 Continued

Sustainability dimension and coding item		*Property (date of management plan)*			
Key	Evident/somewhat evident Not evident	*Florence (2016)*	*Lamu (2013)*	*Melaka (2008)*	*Ahmadabad (2016)*
Values and attitudes	21 Local values and attitudes identified				
	22 Residents issues identified				
	23 Attitudes to heritage identified				
	24 Quality of life identified				
	25 Vision aligns with community values				
Stakeholder participation	26 Stakeholder relationships addressed				
	27 Government participation				
	28 Government influence				
	29 Non-government participation				
	30 Non-government influence				
	31 Local participation				
	32 Local influence				
	33 Visitor participation				
	34 Visitor influence				

government and non-government organisation participation in the heritage planning process and influence over operational management. However, local participation was limited to consultation on draft plans rather than meaningful engagement and longer-term influence. Also noted previously, empowerment through ongoing participation is needed to effect sustainable change.

Conclusion

The aim of this chapter has been to examine the challenges facing historic urban cultural landscapes where balancing socially sustainable development must also consider conserving heritage significance. The chapter has argued for a definition of urban heritage that recognises the complex layers of physical components and patterns, as well as the sociocultural values and traditions that together give rise to a sense of identity and place. The term 'historic urban cultural landscape' has been proposed to reflect a dynamic sociocultural setting rather than the static physical canvas implied by the term 'historic urban environment' (Gibson, 2009; Taylor, 2015). The chapter has also argued that the significance of many historic urban cultural landscapes lies not in their creation as a singular vision or collection of monuments, but in their accumulated layers of development, use, and meaning (Dovey, 2016). An additional thread in the discussion has been the growing recognition that urban heritage can no longer be managed in isolation from the prevailing forces acting elsewhere on a city (Bandarin, 2015). These complex issues can have specific implications for socially sustainable urban heritage, particularly in landscapes that support significant tourism activity.

A framework to manage socially sustainable urban heritage that exhibits three characteristics has been suggested: first, a capacity to establish a specific statement of heritage significance that captures those tangible and intangible characteristics valued by a broad selection of stakeholders. This generates a negotiated understanding of a collective identity and sense of place, and works to establish limits of acceptable change for sustainable development. Second is a governance model that embraces a holistic strategic orientation and ongoing stakeholder participation. This ensures that sustainable heritage management goals are supported through consideration of a long-term circular model of causality, although democratic participation is necessary for equitable fairness, empowerment, and accountability. The third characteristic is a set of conservation principles that respect change as an inherent tradition of the city. This aims to support social inclusion and diversity, as well as continued improvements in quality of life. Evidence indicates concerted international efforts to transform the principles that guide the management of historic urban landscapes into a practical framework that accommodates these three key characteristics. These efforts are complicated, however, by the particular challenges posed by social sustainability (Colantonio and Dixon, 2011; Shirazi and Keivani, 2017) and heritage appropriation (Graham *et al.*, 2000; Harrison, 2010; Smith, 2006).

This study of the extent that the principles of socially sustainable development have been integrated into management practices at four World Heritage Listed urban cultural landscapes suggests that, despite recent efforts, five key concerns remain. First, concern surrounds the extent to which the situation analysis in each case is focused on property-specific factors and tangible heritage fabric at the expense of external trends and issues, and intangible cultural practices. Second, an important finding is the limited assessment of local enterprise skills and business capabilities. References were made to traditional tradespeople as an asset and an active tradition of volunteering. However, a socially sustainable approach would aim to strengthen economic vitality, social diversity, and community networks through enterprise development. This deficiency may relate to a lack of appropriate expertise among heritage professionals and the limited integration of World Heritage properties into broader sustainable development policy and planning frameworks. The effect, however, is to isolate properties and reduce their capacity to deliver equitable benefits and self-empowerment.

Third, of particular concern for social sustainability is the persistence of limited community participation. Although government and non-government agencies were included in the planning process and ongoing management of properties, only one management plan indicated that community consultation had taken place and then only in the planning stage. Efforts had otherwise been made to inform and seek opinion, but through consultation rather than any structured delegation of power. This weakness might stem from the difficulty of achieving an equitable participatory process, as well as the danger such a process might pose to traditional authorities. This is a lost opportunity, particularly given that strengthening the appreciation of heritage is a World Heritage objective. The active involvement of a broad selection of stakeholders also engenders a collective sense of ownership and commitment not found through consultation on already developed concepts and strategies.

Fourth, an additional concern for social sustainability and urban heritage is the continual challenge to develop adequate performance indicators. The completion of a specific conservation project or the measurement of visitor demographics is a tangible outcome and relatively easy to measure. The impact of World Heritage status on equitable fairness, inclusion, diversity, and quality of life are intangible and hard to measure outcomes that were absent in all plans. Although this deficiency might also relate to a lack of expertise and the limited integration of properties into broader policy and planning frameworks, the development of uniform indicators would none-the-less support goal-setting and benchmarking from one year to the next, and across properties.

Finally, it should be noted that there are several limitations to this study. The use of a mixed-methods research strategy would provide a more comprehensive understanding of the planning process in each case, and the use of more than one coder would reduce the potential for bias in the analysis. The varied structure of the key documents also affects the validity of cross-case comparison. Despite this, the implications for socially sustainable urban heritage are clear. At a policy level, further work could be done to develop mandatory requirements for socially sustainable development within World Heritage properties. At a practical level, strategic planning needs to be more holistic, stakeholder engagement needs to be more meaningful, and sustainable development objectives need to be more actively amalgamated into planning, implementation and evaluation processes. This suggests a larger role for UNESCO in situational analysis data collection and strategic planning capacity building, the development of uniform approaches to the measurement of heritage listing impacts on social sustainability, and greater technical support to enable the analysis of such data in terms of economic, environmental and social costs, and benefits.

References

Ahmadabad Municipal Corporation (2016) Historic city of Ahmadabad heritage management plan. Report. Available at: file:///Users/uqclando/Downloads/1551-2144-Management%20Plan-en.pdf (accessed 18 October 2017).

Anonymous (n.d.) Lamu Old Town World Heritage Site management plan 2013–2017. Report. Available at: file:///Users/uqclando/Downloads/mgmt-1055-2013%20(1).pdf (accessed 5 October 2017).

Anonymous (2008) Conservation Management Plan for the Historic City of Melaka. Report. Available at: file:///Users/uqclando/Downloads/Mgmt1223-2008-Melaka.pdf (accessed 18 October 2017).

Ashworth G and Tunbridge J (1996) *Dissonant Heritage: The Management of the Past as a Resource in Conflict*. Chichester: Wiley.

Australia ICOMOS (2013) *The Burra Charter: The Australia ICOMOS Charter for Places of Cultural Significance*. Burwood: Australia ICOMOS.

Bandarin F (2015) Urban conservation and the end of planning. In: Bandarin F and van Oers R (eds), *Reconnecting the City: The Historic Urban Landscape Approach and the Future of Urban Heritage*. Chichester: Wiley Blackwell, pp. 1–16.

Bandarin F and van Oers R (2012) *The Historic Urban Landscape: Managing Heritage in an Urban Century*. Chichester: Wiley Blackwell.

Bandarin F and van Oers R (eds) (2015) *Reconnecting the City: The Historic Urban Landscape Approach and the Future of Urban Heritage*. Chichester: Wiley Blackwell.

Blandford C (2006) Management plans for UK World Heritage sites: Evolution, lessons and good practice. *Landscape Research* 31(4): 355–362.

Chhabra D, Healy R, and Sills E (2003) Staged authenticity and heritage tourism. *Annals of Tourism Research* 30(3): 702–719.

Cleere H (1996) The concept of 'outstanding universal value' in the World Heritage convention. *Conservation and Management of Archaeological Sites* 1(4), 227–233.

Colantonio A and Dixon T (2011) *Urban Regeneration and Social Sustainability: Best Practice from European Cities*. Chichester: Wiley Blackwell.

Dalmas L, Geronimi V, Noel J-F, and Tsang King Sang J (2015) Economic evaluation of urban heritage: An inclusive approach under a sustainability perspective. *Journal of Cultural Heritage* 16(5): 681–687.

Darlow S, Essex S, and Brayshay M (2012) Sustainable heritage management practices at visited heritage sites in Devon and Cornwall. *Journal of Heritage Tourism* 7(3): 219–237.

Dovey K (2016) *Urban Design Thinking: A Conceptual Toolkit*. London: Bloomsbury.

Forrest C (2010) *International Law and the Protection of Cultural Heritage*. London: Routledge.

Francini C, Bocchio C, Ippolito V, and Marin M (2016) The Management Plan of the Historic Centre of Florence. Available at: www.firenzepatrimoniomondiale.it/wp-content/uploads/2015/12/Piano-gestione-en-web.pdf (accessed 14 September 2017).

Gibson L (2009) Cultural landscapes and identity. In: Gibson L and Pendlebury J (eds), *Valuing Historic Environments*. Farnham, Surrey: Ashgate Publishing Ltd, pp. 67–92.

Gibson L and Pendlebury J (2009) Introduction: Valuing historic environments. In: Gibson L and Pendlebury J (eds), *Valuing Historic Environments*. Farnham, Surrey: Ashgate Publishing Ltd, pp. 1–16.

Graham B, Ashworth GJ, and Tunbridge JE (2000) *A Geography of Heritage: Power, Culture and Economy*. London: Arnold.

Gullino P, Beccaro GL, and Larcher F (2015) Assessing and monitoring the sustainability in rural World Heritage Sites. *Sustainability* 7(14): 86–210.

Guzman PC, Pereira Roders A, and Colenbrander BJF (2017) Measuring links between cultural heritage management and sustainable urban development: An overview of global monitoring tools. *Cities* 60: 19–201.

Halbwachs M (1926) 1992. *On Collective Memory*. Reprint, Chicago, IL: University of Chicago Press.

Harrison R (ed.) (2010) *Understanding the Politics of Heritage*. Manchester: Manchester University Press.

Harrison R (2013) Forgetting to remember, remembering to forget: Late modern practices, sustainability and the 'crisis' of accumulation of the past. *International Journal of Heritage Studies* 19(6): 579–595.

Hobsbawm E and Ranger T (eds) (1983) *The Invention of Tradition*. Cambridge: Cambridge University Press.

Hsieh H and Shannon S (2005) Three approaches to qualitative content analysis. *Qualitative Health Research* 15(9): 1277–1288.

ICOMOS (1982) Historic Centre of Florence, No. 174 advisory body evaluation. Paris: ICOMOS. Available at: file:///Users/uqclando/Downloads/174-ICOMOS-193%20(3).pdf (accessed 14 December 2017).

ICOMOS (2001) Lamu Old Town, No. 1055 advisory body evaluation. Paris: ICOMOS. Available at: file:///Users/uqclando/Downloads/1055-ICOMOS-1229-en.pdf (accessed 14 December 2017).

ICOMOS (2008) Melaka and George Town, historic cities of the Straits of Malacca, No. 1223 advisory body evaluation. Paris: ICOMOS. Available at: file:///Users/uqclando/Downloads/1223-ICOMOS-1544-en.pdf (accessed 14 December 2017).

ICOMOS (2014) *The Florence Declaration on Heritage and Landscape as Human Values*. Paris: ICOMOS. Available at: http://icomos.org.tr/Dosyalar/ICOMOSTR_0481066001459594024.pdf (accessed 4 March 2018).

ICOMOS (2017) Historic city of Ahmadabad, No. 1551 advisory body evaluation. Paris: ICOMOS. Available at: file:///Users/uqclando/Downloads/1551-ICOMOS-2144-en%20(3).pdf (14 December 2017).

Jessop B (2002) Liberalism, neoliberalism, and urban governance: A state-theoretical perspective. *Antipode* 34(3): 452–472.

Klopp JM and Petretta DL (2017) The urban sustainable development goal: Indicators, complexity and the politics of measuring cities. *Cities* 63: 92–97.

Labadi S and Logan W (2016) Introduction: Valuing historic environments. In: Labadi S and Logan W (eds), *Urban Heritage, Development and Sustainability: International Frameworks, National and Local Governance*. Abingdon, Oxon: Routledge, pp. 1–16.

Landorf C (2009a) A framework for sustainable heritage management: A study of UK industrial heritage sites. *International Journal of Heritage Studies* 15(6): 494–510.

Landorf C (2009b) Managing for sustainable tourism: A review of six cultural World Heritage Sites. *Journal of Sustainable Tourism* 15(6): 53–70.

Landorf C (2011) Evaluating social sustainability in historic urban environments. *International Journal of Heritage Studies* 17(5): 463–477.

Licciardi G and Amirtahmasebi R (eds) (2012) *The Economics of Uniqueness: Investing in Historic City Cores and Cultural Heritage Assets for Sustainable Development*. Washington, DC: The World Bank.

Lowenthal D (1998) *The Heritage Crusade and the Spoils of History*. Cambridge: Cambridge University Press.

Malcolm-Davies J (2004) Borrowed robes: The educational value of costumed interpretation at historic sites. *International Journal of Heritage Studies* 10(3): 277–293.

Poulios I (2014) Discussing strategy in heritage conservation: Living heritage approach as an example of strategic innovation. *Journal of Cultural Heritage Management and Sustainable Development* 4(1): 16–34.

Rey-Perez J and Avila MES (2017) Historic urban landscape: An approach for sustainable management in Cuenca (Ecquador). *Journal of Cultural Heritage Management and Sustainable Development* 7(3): 308–327.

Rodwell D (2002) The World Heritage Convention and the exemplary management of complex heritage sites. *Journal of Architectural Conservation* 8(3): 1–21.

Rodwell D (2007) *Conservation and Sustainability in Historic Cities*. Oxford: Blackwell Publishing.

Shirazi R and Keivani R (2017) Critical reflections on the theory and practice of social sustainability in the built environment – A meta-analysis. *Local Environment* 22(12): 1526–1545.

Smith L (2006) *Uses of Heritage*. Abingdon, Oxon: Routledge.

Sowinska-Swierkosz B (2017) Review of cultural heritage indicators related to landscape: Types, categorisation schemes and their usefulness in quality assessment. *Ecological Indicators* 81: 526–542.

Stovel H (1994) Notes on authenticity. In: Larsen KE and Marsten N (eds), *Preparatory Workshop, Conference on Authenticity in Relation to the World Heritage Convention*. Bergen: Tapir Forlag, pp. 125–198.

Taylor K (2015) Cities as cultural landscapes. In: Bandarin F and van Oers R (eds), *Reconnecting the City: The Historic Urban Landscape Approach and the Future of Urban Heritage*. Chichester: Wiley Blackwell, pp. 179–202.

Throsby D (2016) Investment in urban heritage conservation in developing countries: Concepts, methods and data. *City, Culture and Society* 7(2): 81–86.

Throsby D (2010) *The Economics of Cultural Policy*. Cambridge: Cambridge University Press.

UNESCO (1972) *Convention Concerning the Protection of the World Cultural and Natural Heritage*. Paris: UNESCO. Available at: https://whc.unesco.org/en/convention-text/ (accessed 27 February 2018).

UNESCO (2002) *Budapest Declaration on World Heritage*. Paris: UNESCO. Available at: https://whc.unesco.org/en/decisions/1217/ (accessed 27 February 2018).

UNESCO (2005) *Vienna Memorandum on 'World Heritage and Contemporary Architecture – Managing the Historic Urban Landscape'*. Paris: UNESCO. Available at: http://unesco.lt/uploads/file/failai_VEIKLA/kultura/RA_Vienna_memorandum.pdf (accessed 27 February 2018).

UNESCO (2011) *Recommendation on the Historic Urban Landscape*. Paris: UNESCO. Available at: https://whc.unesco.org/uploads/activities/documents/activity-638-98.pdf (accessed 27 February 2018).

UNESCO (2013) *The Hangzhou Declaration: Placing Culture at the Heart of Sustainable Development Policies*. Paris: UNESCO. Available at: www.unesco.org/new/en/culture/themes/culture-and-development/hangzhou-congress (accessed 27 February 2018).

UNESCO (2015) *Operational Guidelines for the Implementation of the World Heritage Convention*. Paris: UNESCO. Available at: http://whc.unesco.org/archive/opguide12-en.pdf (accessed 4 March 2018).

United Nations (2014) *World Urbanization Prospects: The 2014 Revision – Highlights*. Available at: https://esa.un.org/unpd/wup/publications/files/wup2014-highlights.pdf (accessed 14 December 2017).

United Nations (2015) *Transforming Our World: The 2030 Agenda for Sustainable Development*. Available at: www.un.org/ga/search/view_doc.asp?symbol=A/RES/70/1&Lang=E (accessed 27 February 2018).

Urry J (1995) *Consuming Places*. London: Routledge.

van Oers R and Pereira Roders A (2017) Aligning agendas for sustainable development in the post 2015 world. *Journal of Cultural Heritage Management and Sustainable Development* 4(2): 122–132.

Wilson L and Boyle E (2006) Interorganisational collaboration at UK World Heritage Sites. *Leadership and Organisational Development Journal* 27(6): 501–523.

Wright WCC and Eppink F (2016) Drivers of heritage value: A meta-analysis of monetary valuation studies of cultural heritage. *Ecological Economics* 130: 277–284.

Yung HKE and Chan HWE (2012) Critical social sustainability factors if urban conservation: The case of the central police station compound in Hong Kong. *Facilities* 30(9/10): 396–416.

6 Spatiality of social sustainability

Social activity and neighbourhood space

M. Reza Shirazi and Ramin Keivani

On the 'spatial' in social sustainability discourse

In Chapter 1, we listed a number of definitions for social sustainability. These are based on a spatial element that either addresses the 'place' within which social sustainability themes and qualities, regardless of how they are defined, are positioned, or signifies a 'sphere' across which these qualities are individually and collectively perceived and comprehended. This spatiality, however, is differently implied. Approaching the city as a 'long-term viable setting' that can function as a 'viable urban social unit' (Yiftachel and Hedgcock, 1993: 140) highlights the capacity of a city to be a locality for delivering social qualities. The search for 'equitable access to urban opportunities' (Boschmann and Kwan, 2008: 139) has spatial implications, because the underlying prerequisite to achieve this goal is partly related to urban space and fair distribution of urban facilities across the city. From another perspective, 'harmonious evolution of civil society' (Stren and Polese, 2000: 15), a 'decent quality of life or livelihood' (Koning, 2002: 70), and 'well-being of the people' (Søholt *et al.*, 2012: 256) hint at a subjectivity that goes beyond the locality as a physical setting and refers to a collective imagination in the minds of the urbanites. Despite these indirect references and spatial implications, the spatiality of social sustainability remains largely unexplored and untouched.

Overall, the literature suggests that social sustainability qualities have two natures: hard qualities that are primarily physical and tangible, such as equal access to urban services and quality of neighbourhood, and soft qualities that are essentially non-physical and intangible, such as social interaction and participation (Shirazi and Keivani, 2017). These two sets of qualities are spatial in different ways: although hard qualities more explicitly address physicality of the 'place' and access to urban objects, soft qualities more implicitly refer to the physical/non-physical 'sphere' of urban practices and sociocultural manifestations. For example, equal access to urban facilities necessitates positioning of basic urban services within the accessible vicinity to all citizens. Spatiality of social interaction, on the other hand, underpins both physical places of socialising from street corners to plazas, and communicational spheres of social exchanges and realms of collective perceptions and imaginations, which are partly practised in the virtual space of social media. In this sense, social sustainability is implicitly and explicitly spatial: it simultaneously and equally underlines the 'spatiality of place' and the 'sociality of space'.

From another point of view, social sustainability is about the perception of the inhabitants with regard to the qualities of social sustainability. A socially sustainable environment is the one that is perceived by the residents as socially sustainable, an environment in which essential qualities of social sustainability are perceived and comprehended to be of a high standard. This perception, although rooted in the individual's mind, is constructed as the result of the aggregation of individuals' perceptions incorporated into a collective imagination. This dimension derives from the very fact that higher social sustainability is not necessarily associated with the locality of higher quantifiable living standards and well-equipped built environments, but with the collective imagination of urbanites and their perception of that locality. This understanding pinpoints a different dimension of the spatiality of social sustainability: intersubjective spatiality. Social sustainability of a given place, in this sense, is the construction of collective subjective perception of the urbanites (inhabitants) from essential, physical/non-physical qualities of social sustainability. This intersubjectivity is a mental sphere horizontally expanded over the urban space and beyond, and vertically inclusive of all urbanites regardless of sex, social status, age, and ethnicity. This intersubjectivity covers a range of issues such as satisfaction with physicality of the environment, through social structures, relations, and practices, to feelings of safety and attachment. It is a collective imagination.

Thus, we would argue, spatiality of social sustainability is constructed through objective spatiality of physical space as well as intersubjective spatiality of individuals' and urbanites' perception. Objective spatiality is about the 'empirical space', in that it is defined by dimensional measurements and trigonometric descriptions of the geometrical relationships between objects, based on an empirico-physical approach to space (Shields, 1997). Intersubjective spatiality, on the other hand, refers to the collective field of subjective spaces created as the aggregation of the perceptual imagination of individuals. Although these two modes of spatiality overlap and are interwoven, they take place in two realms: the former in the tangible locality of urban space and the latter in the mental spatiality of human minds. As Table 6.1 shows, these two modes of spatiality of social sustainability are associated with two theoretical grounds: although objective spatiality is essentially about the theory of urban form, intersubjective spatiality is related to the social theory of space.

The theory of urban form is not a cohesive body of knowledge; it refers to multiple scholarships primarily developed in different disciplines such as architecture, urban history, geography, and urban design, but interacts with other disciplines such as urban sociology, urban theory, and theory of space. Generally speaking, theory of urban form gives priority to the material dimension of the built environment and the morphological deployment of urban components, though social aspects are explicitly or implicitly incorporated. For

Table 6.1 Two modes of spatiality of social sustainability

	Modes of spatiality	*Theoretical foundation*	*Relevant disciplines*	*Methods for study*
Spatiality of social sustainability	Objective spatiality	Theory of urban form	Architecture, geography, urban design, urban history, urban morphology, etc.	Predominantly quantitative
	Intersubjective spatiality	Social theory of space	Urban sociology, urban geography, urban theory, theory of space, etc.	Predominantly qualitative

example, the history of urban form investigates different types of city development models, evolution of urban structure, and morphological arrangement of urban components (Benevolo, 1980; Kostof, 1991; Morris, 1994). A branch of geography employs an empirical–analytical approach to the city in order to document spatial organisation of urban form (Herbert and Thomas, 1982). Space syntax theory, developed by Bill Hillier and Julienne Hanson, investigates configurational properties of urban structure by measuring the relationship between urban components and syntactical measurement of street layouts (Hillier, 1996, 1999, 2002; Hillier and Hanson, 1989; Hillier *et al.*, 1976, 1993). Urban morphology, with its three main English, Italian, and French branches, is another school of studying urban form that investigates evolution of cities from the beginning up to the present and transformation of various urban components such as buildings, gardens, streets, parks, etc. (Clark, 1985; Goodall, 1987; Moudon, 1997; Small *et al.*, 2001). Other key approaches to urban form, sometimes from a non-physical perspective, include mental mapping and perceptual form (Lynch, 1960), cultural and behavioural values (Rapoport, 1977), and pattern language of architectural and urban forms (Alexander, 1977), to name but a few.

Social theory of space originated from social science but expanded to urban studies, urban planning, and the built environment. As Castells puts it, 'There is no theory of space that is not an integral part of a general social theory' (Castells, 1977: 115). Marxist and Weberian urban sociology, urban sociology of the Chicago School, and social theory of the Frankfurt School are only a number of schools of thought that have contributed to the social theory of space. To this we should add many individuals who contributed to the expansion and enrichment of social theory of space. All these schools and individuals deal with questions that are about the social aspects of space, or sociality of space, and thus address the underlying logic of intersubjective spatiality of social sustainability, such as spatial relations and forms of sociation (*Vergeselschaftung*) (Simmel, 2004), social problems (Burgess, 1925), social relations (Park, 1925), public sphere (Habermas, 1989), social spatialisation (Shields, 1991), habitus and social space (Bourdieu, 1977, 2010), socio-spatial dialectic (Soja, 1989), social justice (Harvey, 2009), space of flows (Castells, 2005), perceived and lived space (Lefebvre, 1991), and subjectivity in the city (Tonkiss, 2005). To be sure, this list is not exhaustive, but shows the complexity of the subject and theoretical foundations of the intersubjective spatiality of social sustainability.

These two modes of spatiality demand different epistemological approaches: although studies about theory of urban form employ predominantly, but not exclusively, quantitative methodologies and tools, investigations on the social theory of space advocate predominantly qualitative methods and inquiry techniques. In other words, as objective spatiality of social sustainability mainly takes place in the materiality of the urban space, it is observable and thus quantitatively measurable. Intersubjective spatiality of social sustainability, on the other hand, occurs in the non-physical space of collective minds, and hence is not observable but perceivable and comprehensible through advanced social and ethnographic methodologies and tools. Despite these methodological preferences, mixed methodologies are intensively used, in both realms, to achieve higher reliability and accuracy. It should be noted that the division of objective and intersubjective spatiality is not an ontological divide, but a methodological and epistemological one. In fact, these two realms are essentially interlinked and ontologically interwoven, as most of the thinkers we have referred to here have already underlined. In the end, a holistic understanding should overlap these two modes of spatiality to construct an integrated image. As noted later, in this chapter we mainly concentrate on the objective spatiality of social sustainability.

Mapping outdoor social activity

Urban space, from a block to the entire city, is the domain of social actions and activities. Outside our immediate private space, called outdoor space, is the place of our individual and collective activities, which differ from indoor activities in that outdoor activities are observable and watchable by the public, and involve people other than our family members. The outdoor space registers itself as heterogeneous from two perspectives: the 'place' of activity and the 'type' of activity. In this sense, outdoor space consists of an agglomeration of heterogeneous zones of different activities with social implications. The heterogeneity of outdoor space is determined by the spatial configuration of the environment, rooted in different characteristics of urban form such as land use, urban layout, street pattern, building typology, etc. In fact, places accommodate different activities in terms of type and intensity, because they have different spatial configurations: areas that accommodate and encourage moving are different in spatial configuration and land use from areas that house and promote interaction and communication. Correlation between outdoor social activity and urban form is not straightforward, but necessitates a deep knowledge of the following key questions: What are the main outdoor social activities, and what is their intensity? How are outdoor social activities spatially distributed and practised? How are these activities correlated with the urban form characters such as land use and street layout? Are outdoor social activities gender and age determined? What is the social logic of spatial distribution of activities? How is the pattern of these activities temporal, and what is the reason behind it? All these questions address the objective spatiality of social sustainability as discussed earlier.

In this chapter, we introduce a methodology to measure the objective spatiality of social sustainability by which the above-mentioned questions could be answered. We apply this method to the case of the Bethnal Green neighbourhood in London and discuss different dimensions and aspects of the spatiality of social sustainability.

Methodology

To map outdoor social activities, we need to take into consideration the following critical elements: type of activities, age and gender of actor, pattern of activity, place of activity, and time of activity. We use the label 'incident' to refer to a person of particular gender and age carrying out an activity of any type. To map these activities, we record each incident based on three components: type of activity, gender of actor, and age of actor. Each incident, or person activity, is, thus, coded as a three-component code of XYZ: X shows the activity type, Y refers to gender, and Z indicates the age of the individual. The main activities that are normally observable in the outdoor space could be categorised into the following types: moving (walking), sitting, standing, conversation, playing, group playing, working (including gardening, fixing a car, cleaning front door, etc.), cycling, and running (e.g. jogging). Each activity type is coded differently, for example M stands for moving and S for sitting. Gender is recorded as F for female and M for male. Four categories of age include children (roughly up to 9 years old), teenagers (roughly 10–17), adults (18 and over), and elderly people (those over 70). Thus, a moving elderly female is coded as MFE and a cycling teenage girl as YFT.

Observation was conducted on a weekday (Tuesday, Wednesday, or Thursday) and at the weekend (Saturday). Monday and Friday were excluded because, as the first and last weekdays, they normally have different requirements which may have an impact on residents' interactions and activity pattern. To avoid any climatic distraction, observation was conducted on sunny days, in a non-winter period, with an average temperature of 18–22°C. The observer followed a predefined path, which covered all the major routes and outdoor spaces of the area under investigation. A complete round of observation took place in each time-slot of 8 a.m.–10 a.m., 10 a.m.–12 p.m., 12 p.m.–2 p.m., 2 p.m.–4 p.m., and 4 p.m.–6 p.m. Incidents observable in the immediate surrounding of the observer were mapped using the defined codes and symbols. Printed maps that follow the predefined route order are available for each time-slot observation. Incidents are noted in their precise locations on the map as they occurred on the ground.

For data processing and analysis, ArcMap software was used. Data was added as a feature to the existing geographical information system (GIS) maps. This enabled running numerous and complex inquiries based on activity type, gender, and age on both weekdays and weekends. It also enabled us to overlap produced maps with other maps such as land use map.

Bethnal Green study area, overview

The study area is bounded by Hackney Road (north), Cambridge Heath Road (east), Bethnal Green Road (south), and Warner Place and Squirries Street (west), and is part of the London Borough of Tower Hamlets, Bethnal Green North Ward (Figure 6.1). Originally a semi-rural parish in 1743, the area was extensively developed in mid-nineteenth century, but was reported as the poorest parish of London in 1871, suffering from low-quality buildings,

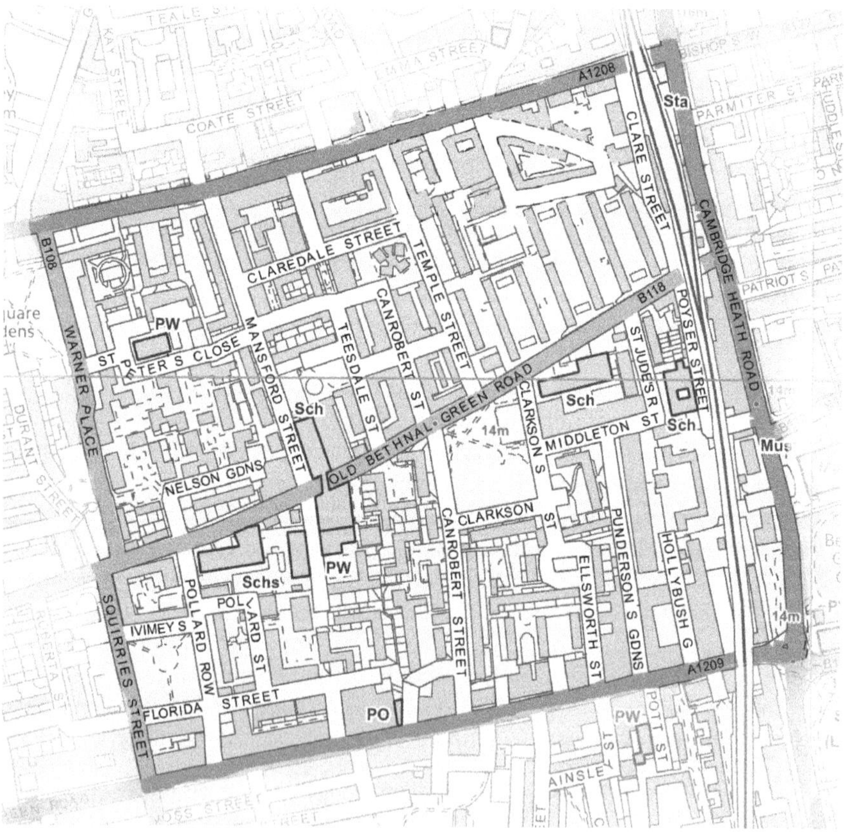

Figure 6.1 Bethnal Green study area.

poor urban services, and an overcrowded and dilapidated environment. Becoming part of the County of London in 1889, Bethnal Green accommodated displaced people from all over Britain and the world. This resulted in population growth (129,727 in 1901), high population density (420 people per hectare), and poverty and deprivation (Porter, 1996). Degradation and deprivation remained a crucial issue during the early twentieth century, encouraging the London County Council to implement an intensive clearance scheme during the 1920s and 1930s (Baker, 1998).

After the extensive damage of World War II, slum clearance and construction of new estates changed the urban landscape to a modern environment and improved living standards (Butler and Rustin, 1996). This turned Bethnal Green into a homogeneous community of mainly English working class with strong local loyalties, family ties, and a matriarchal family structure (Young and Willmott, 1957). Starting from the 1980s, Bethnal Green diversified socially and culturally (Rix, 1996). A main reason for this was the immigration influx of the 1970s and 1980s, particularly Pakistanis and later Bangladeshis (Dench *et al.*, 2006; Pollen, 2002), so that from the 1990s Bethnal Green presented itself as a cosmopolitan, multicultural, and multi-ethnic area (Dench *et al.*, 2006). However, different types of deprivation, such as income and crime, are still a critical challenge.

The study area has a population of around 6,815, accommodated in 2,620 dwellings. The population density is 190.8 pph (person per hectare), which is higher than the average in Bethnal Green North Ward (166.3 pph). About half of the inhabitants (48.9%) are from different white backgrounds and around 33% are Bangladeshi. Of the inhabitants 34.3% are considered to be Muslims, 24.7% as Christians. Buddhists, Hindus, and Jewish people are only about 0.6% of the population each.

In terms of land use, only 17% of the plots located within the study area are mixed use (Figure 6.2), and these are concentrated along the Bethnal Green Road, Cambridge Heath Road, Hackney Road, as well as Winkley Street at the heart of the neighbourhood. The inner zones of the area accommodate a number of residential estates built in different time periods after the World War II. The study area is rich in terms of public green spaces, parks, and community gardens. At the heart of the area is the Middleton Green Park (approximately 8,700 m^2), with playground and basic outdoor sport facilities for children and teenagers. The study area is well connected to the city through surrounding thoroughfares; public transportation is available within walking distance including overground, underground, and bus services. There is no cycling path across the streets. Space syntax analysis of the area suggests a relatively high integration with the entire city (global integration), but the estates present themselves as the least integrated at the local scale. In other words, spaces within the estates are spatially the least connected and highest segregated within the study area. The analysis also suggests that the Old Bethnal Green is a significant movement axis at the local scale.

Figure 6.2 Mixed land use plan, Bethnal Green – darker plots are mixed use.

Different building typologies are observable, from Victorian terraced houses to the post-war developments and high-rises. The majority of buildings are flats (88%). Of the buildings 34% are three-storey, 20% two-storey, and 18% four-storey in height. Although the intersection of the Old Bethnal Green Road and Canrobert Street can be considered as the heart of the neighbourhood, enriched by the Middleton Green with its greenery and playgrounds, Bethnal Green Road is the busiest urban axis and accommodates a wide range of shops, offices, and urban services.

Pattern of outdoor social activities

In the Bethnal Green area, the total number of incidents recorded on both weekdays and weekends was 6,075, of which 3,042 cases took place on weekdays and 3,033 cases at the weekend. This implies that there is no significant difference between activity numbers during the weekdays and weekends, so that neighbourhood space is equally used throughout the week.

Typological pattern of activities

Typological pattern of activities shows the overall distribution and intensity of different types of activities. As Table 6.2 indicates, moving is by far the dominant activity (66%) followed by standing (12.9%). Other activities comprise less than 10% of the total activities. After moving and sitting, cycling is the most observed activity on weekdays, probably because it is used as a means of transport for commuting between home and workplace. Playing and group playing are higher at the weekend; this is because the school students and children are free at the weekend or can be accompanied by their parents. Conversation as an important type of socialising constitutes only 2.35% of the activities. This suggests that there is either a lack of proper public space for socialising, or people are reluctant to use these spaces. Overall, Table 6.2 suggests that the neighbourhood space is predominantly a space for moving; activities that involve socialising and exchange are only a small part of the outdoor activities.

Temporal pattern of activities

The temporal pattern of activities explores the relationship between different types of activities and time; it indicates how the intensity of activities changes during the day. As Table 6.3 depicts, on weekdays, 8–10 a.m. is the busiest time-slot when inhabitants of different age groups go to their places of activity (work, school, etc.). The figure goes down but rises from the noon onwards and reaches its zenith between 16 p.m. and 18 p.m. At the weekend, the activity pattern is quite different in the sense that it rises gradually from early morning to afternoon and reaches its peak between 16 p.m. and 18 p.m.

Table 6.3 also illustrates the temporal pattern of each activity type separately. During the weekdays, 8–10 a.m. has the largest number of reported

Table 6.2 Typological pattern of activities

Type of activity	*Number (%)*	*Weekday (%)*	*Weekend (%)*
Moving	4,043 (66)	2,054 (51)	1,989 (49)
Sitting	289 (4.7)	99 (34)	190 (66)
Standing	786 (12.9)	429 (55)	357 (45)
Playing	19 (0.31)	4 (21)	15 (79)
Group playing	85 (1.39)	9 (11)	76 (89)
Conversation	143 (2.35)	68 (48)	75 (52)
Cycling	419 (6.89)	233 (56)	186 (44)
Working	236 (3.88)	124 (53)	112 (47)

Table 6.3 Temporal pattern of activities

		Weekday activities					*Weekend activities*				
Time-slots		8–10	10–12	12–14	14–16	16–18	8–10	10–12	12–14	14–16	16–18
Number of activities		785	481	519	556	701	240	475	696	796	826
Percentage of activities		25	16	17	18	23	7	16	23	26	27
Type of activities	Moving	515	319	379	379	491	154	327	487	478	528
	Sitting	9	12	15	15	31	4	13	36	36	76
	Standing	120	59	61	61	77	31	62	81	81	88
	Playing	1	0	2	0	3	0	0	2	2	7
	Group playing	1	0	1	1	4	0	3	16	16	25
	Conversation	8	14	23	23	11	6	8	4	4	34
	Cycling	93	35	20	20	57	17	37	45	45	43
	Working	26	40	18	18	21	20	16	29	29	21

moving activities, which shows inhabitants travelling to their daily destinations. This figure drops at 10 a.m. but gradually increases towards the evening. At the weekend, however, moving is low in the morning and gradually increases to reach its peak in the afternoon. During both weekdays and weekends, sitting increases gradually, but standing has an irregular pattern. Site observations indicate that most of the standing cases were located around bus stops and behind red lights at pedestrian crossings. Playing and group playing happen more at the weekend, and the number increases towards the afternoon. Conversation mainly takes place in the afternoon. On the weekdays, cycling as a means of transport is used intensively during the early morning and late afternoon. At the weekend, the use of bicycles increases from morning to afternoon.

Gender pattern of activities

Generally speaking, male activities are 16% higher than female activities (Table 6.4). This may reflect the sociocultural structure of the Bethnal Green area (high percentage of Bangladeshis), which may suggest a lower presence of female inhabitants in the public space, and portrays a familial structure with a stronger role for female inhabitants inside their homes. This is supported by Dench *et al.* (2006) who confirm that Bangladeshi community members continue practising their traditions and culture, such as having extended family households, and respecting familial relationships and interest.

Difference in gender pattern of activities is more significant when we consider different types of activities (Table 6.5). For example, group playing is dominated by males whereas females are almost absent. Taking into account that group playing takes place in playing grounds and green spaces, and mainly involves teenagers and children, this may indicate that females do not feel safe or comfortable in these spaces, or social and cultural restrictions do not allow them to play in public areas. Cycling is also a male-dominant activity, the share of female cyclists only being 29% of all cycling cases. Due to the fact that there is no designated cycling path available, this could mean that the lack of safety makes females less interested in cycling in the area. Official statistics indicate that cycling is remarkably gendered across the UK, so that the low level of cycling is not an issue exclusive to Bethnal Green. According to the National Travel Survey (Department for Transport, 2017), during 2014–2016 males aged over five made about three times as many cycle trips

Table 6.4 Gender pattern of activities, weekday and weekend

Gender	*Total number of activities*	*Number of activities/weekday*	*Number of activities/weekend*
Female	2,493 (42%)	1,309 (52%)	1,184 (48%)
Male	3,529 (58%)	1,708 (48%)	1,821 (52%)

Table 6.5 Gender pattern of activities by type

Gender	*Moving*	*Sitting*	*Standing*	*Playing*	*Group playing*	*Conversation*	*Cycling*	*Working*
Female	1,831 (45%)	118 (41%)	317 (40%)	11 (57%)	6 (7%)	35 (24%)	120 (29%)	34 (14%)
Male	2,211 (52%)	171 (59%)	468 (60%)	8 (43%)	71 (83%)	72 (51%)	299 (71%)	202 (86%)
Mix	–	–	–	–	8 (10%)	36 (25%)	–	–

as females (27 as opposed to 9 for females), and cycled around four times as many miles as females (87 as opposed to 20 miles for females). An implication of this gender difference in using neighbourhood space is that urban space could be encoded with ideas of gender (Andrews, 2016; Beebeejaun, 2017; Doan, 2010; Fenster, 2005; Karsten, 2003; Massey, 1994; McDowell, 1983; Scraton and Watson, 1998; Spain, 1992, 1993), and cycling as a social activity could be gendered (Dickinson *et al.*, 2003; Garrard, 2003; Garrard *et al.*, 2008; Pucher and Buehler, 2012; Root and Schintler, 1999; Steinbach *et al.*, 2011).

Age pattern of activities

There is no significant difference among different age groups in terms of the number of activities on weekdays and at the weekend (Table 6.6). Overall, adults are the most present age group, and elderly people the least present. The latter is consistent with the dominant assumption that cities and urban spaces are not well designed and prepared for elderly people in terms of safety, addressing their basic needs, promoting social engagement, and maximising independence and easy movement (Burton and Mitchell, 2006; Fobker and Grotz, 2006; van Den Berg *et al.*, 2015; Wahl and Weisman, 2003), whereas the creation of age-friendly communities is now promoted intensively to give equal opportunities for different age groups (Alley *et al.*, 2007; Lui *et al.*, 2009; Menec *et al.*, 2011; World Health Organization, 2007).

Spatial pattern of outdoor social activities

The spatial pattern of outdoor activities refers to the way in which these activities are spatially distributed across the neighbourhood space in terms of type, gender, and age. This helps us to understand the places of concentration and intensity, and the correlation between activity intensity and characteristics of urban form such as land use and street layout.

Spatial pattern of activity type

Figure 6.3 shows all activities that have taken place within the neighbourhood space during the weekends and weekdays. Each point represents one

Table 6.6 Age pattern of activities

	Total	*Weekday*	*Weekend*
Children	462	240	222
Teenagers	465	247	218
Adults	4,925	2,447	2,478
Elderly people	205	99	106

incident as explained. Figure 6.3 shows that there is a concentration of activities across the main streets. The most crowded axis is the Bethnal Green Road in the south, then Cambridge Heath Road (east). Old Bethnal Green (centre) and Hackney Road also present a relatively high concentration of activities. Some interior secondary streets also present a dense activity pattern, such as Middleton Green at the heart of the study area.

Analysis of the spatial pattern of activity types suggests that the moving pattern clearly corresponds with the street networks. Sitting activity is very limited across the neighbourhood and is concentrated in green areas, bus stops, or in front of cafés and restaurants that offer outdoor services. This indicates the poor condition of urban furniture and the lack of proper communal spaces for sitting and socialising. There are a number of playing grounds in the area that accommodate playing and group playing activities and are accessible for all inhabitants within a reasonable walking distance. The lack of playing activity outside designated playing grounds or within the

Figure 6.3 All activities, all days.

residential areas is mainly due to the fact that the management system of estates has prohibited playing on the public areas within their premises.

As a result of the lack of designated cycling paths, cycling takes place with low intensity across the area, with a concentration on the primary street networks, although the number of female cyclists is far fewer than males, which may reflect the low level of feeling of safety and possibly sociocultural obstacles. This is consistent with the existing literature about the importance of cycling paths and traffic-calming measurement for encouraging cycling activity (Akar and Clifton, 2009; Buehler and Pucher, 2012; Dill and Carr, 2003; Garrard *et al.*, 2008; Nelson and Allen, 1997; Pasha *et al.*, 2016; Wegman *et al.*, 2012; Wheeler, 2013).

The spatial pattern of conversation is quite noticeable. Very limited conversation cases (as noted, only 2.35% of all activities are conversation) are scattered across the area with no significant concentration at specific places. This indicates that densely populated inner areas such as estate residential complexes do not encourage residents to socialise and have face-to-face conversation.

Spatial pattern of gender

Although as previously noted the number of male activities is higher than female activities, comparison of Figures 6.4a and 6.4b shows no significant difference in terms of spatial distribution. In other words, the neighbourhood space is evenly used by males and females. The only noticeable difference is along the Clare Street on the north-east corner beneath the London Overground. This street accommodates workshops and car services, which are male-dominant occupations. Overall, it could be said that there is no spatial segregation based on gender at the neighbourhood scale, and all spaces are equally accessible to the people.

(a) (b)

Figure 6.4 (a) Female activities, all days; (b) male activities, all days.

Spatial pattern of age

Children are homogeneously present across the neighbourhood, with a relative concentration in parks or in the vicinity of schools. Teenagers, interestingly, follow a different distribution pattern to other age groups; they are less interested in the most public and crowded areas such as high streets and the main arteries, but gather in playgrounds and parks. This is plausibly due to having greater freedom for their behaviour and staying out of public scrutiny in such locations. Site observation also confirms that teenagers are boisterous and loud in their collective activities, and this encourages them to gather in places of less visibility so that they feel more comfortable engaging in their games and group activities. Adults are present everywhere and are the dominant actors of the neighbourhood space. Elderly people, as the least present age group, either use the high street for shopping and socialising purposes, or prefer quiet and non-traffic places inside the urban blocks where they feel safer and more comfortable.

Outdoor social activity, land use, and street layout

There is an evident correspondence between outdoor social activity and mixed land use. As Figure 6.5 depicts, activities are concentrated along the streets with dominantly mixed-use buildings. These activities are also denser in predominantly non-residential zones than in predominantly residential ones. In this sense, areas that are predominantly mixed land use and non-residential are places with higher outdoor social activities of any type, gender, and age group. This corresponds with this idea that mixed land use potentially promotes social life in urban space (Brown and Lombard, 2014; Colquhoun, 2004; Desyllas *et al.*, 2003; Duany *et al.*, 2010; Grant, 2002; Leyden 2003; Rowley 1996; Welsh and Farrington, 2009). Overlapping outdoor social activity and space syntax analysis maps also suggest a clear positive correlation between outdoor social activity and urban layout. Streets with higher global and local integrity are an axis of higher social activity concentration.

As the study area is home to a large number of estates and residential complexes, a closer look at two of them helps us to understand how spatial configuration and layout corresponds with outdoor social activity patterns. Minerva Estate (Figure 6.6a) is located on the north-east side of the study area. Officially opened in 1948, it contains 261 flats for 950 inhabitants. Building blocks are laid out north–south, with plenty of open space between them. However, the area is fenced and the open space between blocks is accessible alternately: one open space is a non-accessible green space, the next one accessible for car parking and pedestrians. Mansford Estate is in the west side of the study area, divided into two parts of Mansford North (Figure 6.6b) and Mansford South (Figure 6.6c) by the Old Bethnal Green Road. The estate consists of about 700 homes in 28 blocks, built in various styles from

Figure 6.5 All activities, all days map overlapped with mixed land use plan – darker plots are mixed use

1957 to 1976. Building blocks are diverse in shape, height, and layout, making the complex heterogeneous in spatial configuration. Open spaces between the blocks are mainly dead-end but accessible for car parking and pedestrians, with green space in front of the blocks.

Overall, outdoor social activity within estate complexes follows the spatial logic of the environment. In the Minerva Estate (Figure 6.6a) outdoor activity takes place in open spaces between residential blocks, which are accessible and consequently integrated into the neighbourhood street networks. The southern part of the Mansford Estate (Figure 6.6c) is either empty of any activities, or very crowded because its central space is utilised as an access path – with around 90% of activities as moving – connecting adjacent streets. In this sense, it is the access value of the space and its use for mobility that give it a socially active appearance. The northern part of the Mansford Estate

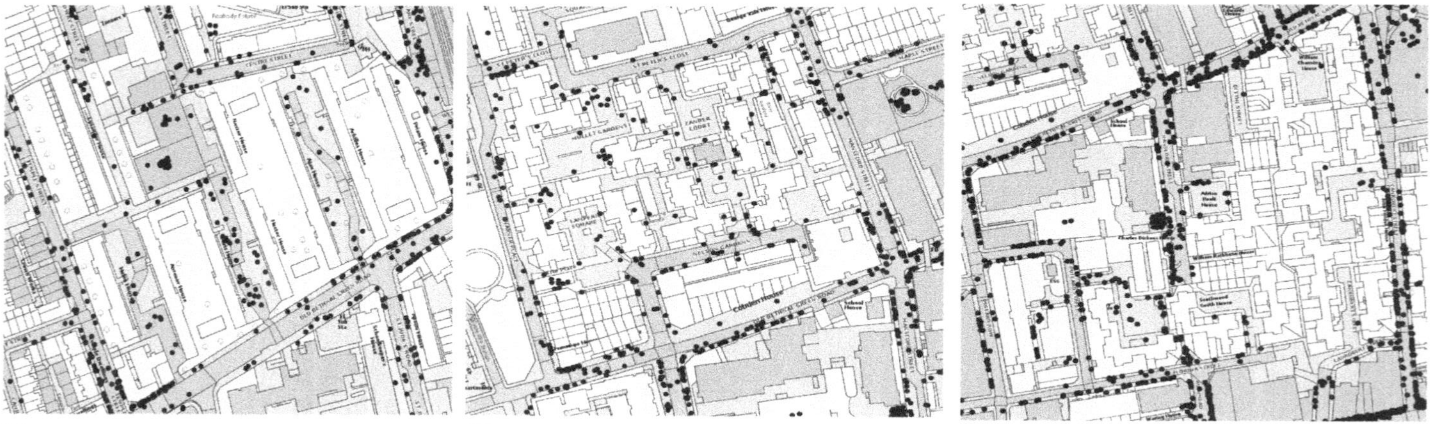

(a) (b) (c)

Figure 6.6 (a) All activities, all days, Minerva Estate; (b) all activities, all days, Mansford Estate North; (c) all activities, all days, Mansford Estate South.

(Figure 6.6b) shows a more homogeneous pattern of social activities, where activities take place in all the spaces between buildings. The complex arrangement of blocks that lacks a spatially privileged axis or route makes the entire space available for outdoor activities, so that, in contrast to the other two cases, different types of activities other than moving take place in different corners of the area.

Conclusion

In this chapter, we have discussed the theoretical foundations of spatiality of social sustainability, introduced a methodology to explore the spatiality of outdoor social activities in urban settings, and investigated a number of spatial questions such as typological, temporal, gender, and age pattern of activities in the case of Bethnal Green. Our theoretical arguments and proposed methodology make significant theoretical, methodological, and practical/policy contributions to the social sustainability discourse.

In terms of theory, we argued that social sustainability has a spatial dimension that has not been discussed and theorised in the literature. We identified two modes of spatiality of social sustainability: objective spatiality of the physical environment and intersubjective spatiality of the collective mind. Although the former positions its grounds on the theory of urban form, the latter is associated with social theory of space. These two modes of spatiality are interwoven and interlinked; however, for epistemological and practical reasons we may approach them separately. Our analysis proved that the characteristics of urban form are significantly associated with outdoor social activities in terms of type of activities, and gender and age of actors. This contributes to the long-standing debates on environmental determinism, the idea that human behaviour is determined by the nature of the geographical environment (Lang, 1987), or architectural determinism, the idea that buildings can have systematic effects on human behaviour, individually or collectively (Hillier, 1996). The core question of this debate is how and to what extent human behaviour is determined by the nature of the geographical environment, and how these two interact and influence each other.

The relationship between human behaviour and physical environment, or between spatial life and spatial design, has registered itself as complex and paradoxical. According to Hanson and Hillier (1987), the relationship between architecture and behaviour, or spatial form and society, has been discussed within the two opposite views of correspondence and non-correspondence; the former believes in independence between social grouping and territorial demarcation, and the latter assumes a great correspondence between them. In fact, the underlying logic of the area-based urban initiatives is the idea that physical improvements for sites of crisis (e.g. deprived neighbourhoods, marginalised communities) and betterment of the physical environment culminate in sociocultural betterment. According to Lang (1987) approaches to behaviour and environment dichotomy fall into four categories: free will approach (no environmental effect), possibilistic approach (environment provides only possibilities for behaviour, individuals can make a choice to participate), probabilistic approach (dealing with probabilities that a particular reaction will occur), and deterministic approach (environment determines desired behaviour). Hillier *et al.* (1987) contend that there is strong empirical evidence to suggest that spatial layout of the built environment generates a potential field of probabilistic co-presence and encounter. Hillier proposes that

> the determinable effects of spatial form on people are both limited and precise. Spatial form, I argue, creates the *field of probable* – though not *all* possible – *encounter and co-presence within which we live and move*; and whether or not it leads to social interaction, this field is *in itself an important sociological and psychological resource.*
>
> (Hillier, 1989: 13; original emphasis)

We can thus talk about an opportunistic approach as a mediatory position between free will and deterministic approaches, which indicates that the spatial layout and physical environment *provide* or *preclude* opportunities for interaction and socialisation, but do not cause such interaction to occur. In other words, spatial layout encourages, invites, and sets the ground for socialising, provided that it is intended to do so, or discourages, rejects, and makes any socialising uncomfortable, provided that it lacks needed quality. Our analysis of Bethnal Green advocates the 'opportunistic approach', because we noticed that, despite spatial opportunities for socialising, for example in the case of Mansford Estate, residents are not using public spaces for social interaction. In this sense, to use Gans's terminology (Gans, 1961), the planner's or designer's task is to make available the opportunity for choice.

In terms of methodology, our approach combines a set of mapping and observation techniques and methods developed in different disciplines (Francis, 1984; Gehl, 2010; Grajewski and Vaughan, 2001; Whyte, 1980), incorporates them into advanced GIS technologies, and thus introduces a combined integrative method for measuring outdoor social activities. Using advanced technology enables us to process a large volume of collected data, run different queries, and thus explore different aspects of the subject in question. This method could be scaled up and down to cover a wide range of scales from single building through urban plazas to city and beyond. The employed method has, however, some limitations. For example, it includes only social activities that are observable in the public and semi-public urban arenas; social activities that take place between neighbours either in private realms or through modern communication tools are excluded. Observation took place at one weekday and one weekend day, which brings to the fore the question of representation and reliability. To be sure, a more comprehensive and reliable image could be obtained if observation would be repeated on different days across the year. Moreover, integrating more qualitative data into the methodology framework, such as interviews and focus group discussions, would lead to a more fine-grained analysis and provide more nuanced reflective understanding. However, this method is enough to provide us with insights into the concentration and pattern of activities based on types, gender, and age.

Our argumentations and proposed methodology also have significant practical implications. Policy makers, planners, and designers always search for initiatives, programmes, and plans to improve social sustainability of cities and communities. These initiatives and programmes should, however, be founded on the empirical evidence from the given site, reflect the reality, and envisage a reachable future. Mapping outdoor social activity, in the way we proposed and implemented, provides policy makers with invaluable knowledge about the spatial pattern of social activities and grants them pointers towards imagining a better future. For example, our analysis suggests that the area around the intersection of Old Bethnal Green Road and Canrobert Street has the capacity to accommodate more diverse urban services and shops, and linked to Middleton Green Park serve as the heart of the neighbourhood.

More importantly, this methodology could be scaled up and applied city-wide and beyond the city to set up what we would like to term a 'topography of social sustainability' map: a detailed and comprehensive map in which social sustainability qualities are layered and could be studied in different spatial units. This interactive map will illustrate the horizontal spatiality of social sustainability, and explore geographies of difference where social sustainability qualities and indicators are unevenly present in different urban localities. The more different the social sustainability qualities from place to place across space, the more uneven the urban social sustainability topography. Topography of social sustainability will, thus, have several significant indications, such as spatial distribution of services and opportunities, geographical difference of living quality, (un)just access to urban privileges, level of deprivation, and pattern of territorial marginalisation.

A comprehensive social sustainability topography map illustrates places of crisis, marginalisation, and discrimination, and thus is informative for policy-making and future planning, in order to develop policies to re-spatialise (diminishing uneven urban social sustainability topography) and finally de-spatialise (achieving even urban social sustainability topography) social sustainability.

Acknowledgements

This work was supported by the European Commission under Marie Curie Intra-European Fellowships Actions; Research Executive Agency [grant no 624794].

References

Akar G and Clifton K (2009) Influence of individual perceptions and bicycle infrastructure on decision to bike. *Transportation Research Record: Journal of the Transportation Research Board* 2140: 165–172.

Alexander C (1977) *A Pattern Language: Towns, Buildings, Construction*. New York: Oxford University Press.

Alley D, Liebig P, Pynoos J, Banerjee T, and Choi IH (2007) Creating elder-friendly communities: Preparations for an aging society. *Journal of Gerontological Social Work* 49(1–2): 1–18.

Andrews H (2016) Gendered space. In: Wong A, Wickramasinghe M, Hoogland R, Naples NA, and Andrews H (eds), *The Wiley Blackwell Encyclopedia of Gender and Sexuality Studies*. Singapore: John Wiley & Sons Ltd, pp. 1–3.

Baker TFT (1998) *A History of the County of Middlesex: Volume 11, Stepney, Bethnal Green*. London: British History Online. Available at: www.british-history.ac.uk/vch/middx/vol. 11 (accessed 11 January 2017).

Beebeejaun Y (2017) Gender, urban space, and the right to everyday life. *Journal of Urban Affairs* 39(3): 323–334.

Benevolo L (1980) *The History of the City*. London: Scholar.

Boschmann EE and Kwan M-P (2008) Toward socially sustainable urban transportation: Progress and potentials. *International Journal of Sustainable Transportation* 2(3): 138–157.

Bourdieu P (1977) *Outline of a Theory of Practice*. Cambridge: Cambridge University Press.

Bourdieu P (2010) *Distinction: A Social Critique of the Judgement of Taste*. London: Routledge.

Brown SC and Lombard J (2014) Neighborhoods and social interaction. In: Cooper CL (ed.), *Wellbeing and the Environment*. Chichester: John Wiley & Sons.

Buehler R and Pucher J (2012) Cycling to work in 90 large American cities: New evidence on the role of bike paths and lanes. *Transportation* 39(2): 409–432.

Burgess EW (1925) The growth of the city: An introduction to a research project. In: Park RE and McKenzie RD (eds), *The City*. Chicago, IL: The University of Chicago Press, pp. 47–62.

Burton E and Mitchell L (2006) *Inclusive Urban Design: Streets for Life*. Amsterdam: Routledge.

Butler T and Rustin M (1996) *Rising in the East?: The Regeneration of East London*. London: Lawrence & Wishart.

Castells M (1977) *The Urban Question: A Marxist Approach*. London: Edward Arnold.

Castells M (2005) Space of flows, space of places: Materials for a theory of urbanism in the information age. In: Sanyal B (ed.), *Comparative Planning Cultures*. London: Routledge.

Clark A (1985) *Longman Dictionary of Geography: Human and Physical*. Harlow: Longman Group.

Colquhoun I (2004) Design out crime: Creating safe and sustainable communities. *Crime Prevention and Community Safety* 6(4): 57–70.

Dench G, Gavron K, and Young M (2006) *The New East End: Kinship, Race and Conflict*. London: Profile Books.

Department for Transport (2017) *National Travel Survey: England 2016*. Available at: www.gov.uk/government/statistical-data-sets/nts06-age-gender-and-modal-breakdown#table-nts0608 (accessed 4 August 2017).

Desyllas J, Connoly P, and Hebbert F (2003) Modelling natural surveillance. *Environment and Planning B: Planning and Design* 30(5): 643–655.

Dickinson JE, Kingham S, Copsey S, and Hougie DJP (2003) Employer travel plans, cycling and gender: Will travel plan measures improve the outlook for cycling to work in the UK? *Transportation Research Part D: Transport and Environment* 8(1): 53–67.

Dill J and Carr T (2003) Bicycle commuting and facilities in major U.S. cities: If you build them, commuters will use them. *Transportation Research Record: Journal of the Transportation Research Board* 1828: 116–123.

Doan PL (2010) The tyranny of gendered spaces: reflections from beyond the gender dichotomy. *Gender, Place & Culture* 17(5): 635–654.

Duany A, Plater-Zyberk E, and Speck J (2010) *Suburban Nation: The Rise of Sprawl and the Decline of the American Dream*. New York: North Point Press.

Fenster T (2005) The right to the gendered city: Different formations of belonging in everyday life. *Journal of Gender Studies* 14(3): 217–231.

Fobker S and Grotz R (2006) Everyday mobility of elderly people in different urban settings: The example of the city of Bonn, Germany. *Urban Studies* 43(1): 99–118.

Francis M (1984) Mapping downtown activity. *Journal of Architectural and Planning Research* 1(1): 21–35.

Gans HJ (1961) Planning and social life: Friendship and neighbor relations in suburban communities. *Journal of the American Institute of Planners* 27(2): 134–140.

Garrard J (2003) Healthy revolutions: Promoting cycling among women. *Health Promotion Journal of Australia* 14(3): 213–215.

Garrard J, Rose G, and Lo SK (2008) Promoting transportation cycling for women: The role of bicycle infrastructure. *Preventive Medicine* 46(1): 55–59.

Gehl J (2010) *Cities for People*. Washington, DC: Island Press.

Goodall B (1987) *The Penguin Dictionary of Human Geography*. Harmondsworth, New York: Penguin Books Ltd.

Grajewski T and Vaughan L (2001) *Space Syntax Observation Procedures Manual*. London: University College London.

Grant J (2002) Mixed use in theory and practice: Canadian experience with implementing a planning principle. *Journal of the American Planning Association* 68(1): 71–84.

Habermas J (1989) *The Structural Transformation of the Public Sphere: An Inquiry into a Category of Bourgeois Society*. Cambridge, MA: MIT Press.

Hanson J and Hillier B (1987) The architecture of community: Some new proposals on the social consequences of architectural and planning decisions. *Architecture and Behaviour* 3(3): 251–273.

Harvey D (2009) *Social Justice and the City*. Athens, London: University of Georgia Press.

Herbert DT and Thomas CJ (1982) *Urban Geography: A First Approach*. Chichester: Wiley.

Hillier B (1989) The architecture of the urban object. *Ekistics* 334–335: 5–21.

Hillier B (1996) *Space is the Machine: A Configurational Theory of Architecture*. Cambridge: Cambridge University Press.

Hillier B (1999) The hidden geometry of deformed grids: Or, why space syntax works, when it looks as though it shouldn't. *Environment and Planning B: Planning and Design* 26(2): 169–191.

Hillier B (2002) A theory of the city as object: Or, how spatial laws mediate the social construction of urban space. *Urban Design International* 7(3–4): 153–179.

Hillier B and Hanson J (1989) *The Social Logic of Space*. Cambridge: Cambridge University Press.

Hillier B, Leaman A, Stansall P, *et al.* (1976) Space syntax. *Environment and Planning B: Planning and Design* 3(2): 147–185.

Hillier B, Burdett R, Peponis J, and Penn A (1987) Creating life: Or, does architecture determine anything? *Architecture and Behaviour* 3(3): 233–250.

Hillier B, Penn A, Hanson J, Grajewski T, and Xu J (1993) Natural movement: Or, configuration and attraction in urban pedestrian movement. *Environment and Planning B: Planning and Design* 20(1): 29–66.

Karsten L (2003) Children's use of public space: The gendered world of the playground. *Childhood* 10(4): 457–473.

Koning J (2002) Social sustainability in a globalizing world: Context, theory and methodology explored. In: van Rinsum H, de Ruijter A, Kazancigil YK, Alagh N, Genov N, Koning J, and Siebers H (eds), *More on MOST: Proceedings of an Expert Meeting*, Amsterdam: UNESCO Centre, pp. 63–89.

Kostof S (1991) *The City Shaped: Urban Patterns and Meanings through History*. London: Thames and Hudson.

Lang JT (1987) *Creating Architectural Theory: The Role of the Behavioral Sciences in Environmental Design*. New York: Van Nostrand Reinhold Co.

Lefebvre H (1991) *The Production of Space*. Oxford: Basil Blackwell.

Leyden KM (2003) Social capital and the built environment: The importance of walkable neighborhoods. *American Journal of Public Health* 93(9): 1546–1551.

Lui C-W, Everingham J-A, Warburton J, Cuthill M, and Bartlett H (2009) What makes a community age-friendly: A review of international literature. *Australasian Journal on Ageing* 28(3): 116–121.

Lynch K (1960) *The Image of the City*. Cambridge, MA: London: MIT Press.

Massey D (1994) *Space, Place and Gender*. Cambridge: Polity.

McDowell L (1983) Towards an understanding of the gender division of urban space. *Environment and Planning D: Society and Space* 1(1): 59–72.

Menec VH, Means R, Keating N, Parkhurst G, and Eales J (2011) Conceptualizing age-friendly communities. *Canadian Journal on Aging/La Revue canadienne du vieillissement* 30(03): 479–493.

Morris AEJ (1994) *History of Urban Form: Before the Industrial Revolutions*. Harlow, New York: Longman.

Moudon AV (1997) Urban morphology as an emerging interdisciplinary field. *Urban Morphology* 1: 3–10.

Nelson A and Allen D (1997) If you build them, commuters will use them: Association between bicycle facilities and bicycle commuting. *Transportation Research Record: Journal of the Transportation Research Board* 1578: 79–83.

Park RE (1925) Community organization and juvenile delinquency. In: Park RE and McKenzie RD (eds), *The City*. Chicago. IL: University of Chicago Press, pp. 99–112.

Pasha M, Rifaat S, Tay R, and de Barros A (2016) Urban design and planning influences on the share of trips taken by cycling. *Journal of Urban Design* 21(4): 471–480.

Pollen R (2002) *Bangladeshi Family Life in Bethnal Green*. London: London School of Economics and Political Science.

Porter R (1996) *London: A Social History*. London: Penguin.

Pucher J and Buehler R (2012) *City Cycling*. Cambridge, MA: MIT Press.

Rapoport A (1977) *Human Aspects of Urban Form: Towards a Man-Environment Approach to Urban Form and Design*. Oxford: Pergamon.

Rix V (1996) Social and demographic change in East London. In: Butler T and Rustin M (eds), *Rising in the East?: The Regeneration of East London*, London: Lawrence & Wishart, pp. 20–60.

Root A and Schintler L (1999) Women, motorization and the environment. *Transportation Research Part D: Transport and Environment* 4(5): 353–355.

Rowley A (1996) Mixed-use development: Ambiguous concept, simplistic analysis and wishful thinking? *Planning Practice and Research* 11(1): 85–96.

Scraton S and Watson B (1998) Gendered cities: women and public leisure space in the 'postmodern city'. *Leisure Studies* 17(2): 123–137.

Shields R (1991) *Places on the Margin: Alternative Geographies of Modernity*. London: Routledge.

Shields R (1997) Spatial stress and resistance: Social meanings of spatialization. In: Benko G and Strohmayer U (eds), *Space and Social Theory: Interpreting Modernity and Postmodernity*. Oxford: Blackwell, pp. 186–202.

Shirazi MR and Keivani R (2017) Critical reflections on the theory and practice of social sustainability in the built environment: a meta-analysis. *Local Environment* 22(12): 1526–1545.

Simmel G (2004) The stranger. In: Jenks C (ed.), *Urban Culture: Critical Concepts in Literary and Cultural Studies*. London, New York: Routledge, pp. 73–77.

Small J, Ross S, and Witherick M (2001) *A Modern Dictionary of Geography*. London, New York: Routledge.

Søholt S, Ruud ME, and Braathen E (2012) A question of social sustainability: Urban interventions in critical neighbourhoods in Portugal and Norway. *Urban Research & Practice* 5(2): 256–272.

Soja EW (1989) *Postmodern Geographies: The Reassertion of Space in Critical Social Theory*. London: Verso.

Spain D (1992) *Gendered Spaces*. Chapel Hill, NC: University North Carolina Press.

Spain D (1993) Gendered spaces and women's status. *Sociological Theory* 11(2): 137.

Steinbach R, Green J, Datta J, *et al.* (2011) Cycling and the city: A case study of how gendered, ethnic and class identities can shape healthy transport choices. *Social Science & Medicine* 72(7): 1123–1130.

Stren R and Polese M (2000) Understanding the new sociocultural dynamics of cities: Comparative urban policy in a global context. In: Polese M and Stren R (eds), *The Social Sustainability of Cities: Diversity and the Management of Change*, Toronto: University of Toronto Press, pp. 13–38.

Tonkiss F (2005) *Space, the City and Social Theory: Social Relations and Urban Forms*. Cambridge: Polity.

van Den Berg P, Kemperman A, de Kleijn B, and Borgers A (2015) Locations that support social activity participation of the aging population. *International Journal of Environmental Research and Public Health* 12(9): 10432–10449.

Wahl H-W and Weisman GD (2003) Environmental gerontology at the beginning of the new millennium: Reflections on its historical, empirical, and theoretical development. *The Gerontologist* 43(5): 616–627.

Wegman F, Zhang F, and Dijkstra A (2012) How to make more cycling good for road safety? *Accident Analysis & Prevention* 44(1): 19–29.

Welsh BC and Farrington DP (2009) *Making Public Places Safer: Surveillance and Crime Prevention*. Oxford: Oxford University Press.

Wheeler SM (2013) *Planning for Sustainability: Creating Livable, Equitable and Ecological Communities.* London, New York: Routledge.

Whyte WH (1980) *The Social Life of Small Urban Spaces.* New York: Project for Public Spaces.

World Health Organization (2007) *Global Age-Friendly Cities: A Guide.* Geneva: WHO.

Yiftachel O and Hedgcock D (1993) Urban social sustainability: The planning of an Australian city. *Cities* 10(2): 139–157.

Young MD and Willmott P (1957) *Family and Kinship in East London.* London: Routledge & Kegan Paul.

7 The social effects of architecture

Built form and social sustainability

Vinicius M. Netto, Júlio Celso Vargas, and Renato T. de Saboya

Introduction

There are several competing definitions of social sustainability. These definitions tend to converge on certain aims like *equity* in the form of access to opportunities and resources as well as in the distribution of positive and negative externalities (McKenzie, 2004), *democracy* geared by access to decision processes, *social interactions* able to foster social capital locally, and in other scales (Bramley and Power, 2009), as well as inclusion and tolerance, *social infrastructure* capable of supporting such forms of cohesion in healthy communities, and the ability of communities to reach levels of *social well-being* (see Anand and Sen, 1994; Opp, 2016; Sen, 2013; Shirazi and Keivani, 2017). In this perspective, 'community' is defined as groups or networks linked by relatively stable social relationships based on something actors have in common, generally a common sense of identity or interests (Marshall, 1998). Local communities are likely to involve settings of face-to-face engagements based on spatial proximity as raw materials for interaction and cohesion (see Goffman, 1972).

Interestingly, connections of this field with a consideration of material conditions rich and detailed enough to incorporate architectural features of the built environment are still to be fully brought to light in the social sustainability debate (compare Woodcraft *et al.*, 2012; Opp, 2016). To be sure, aspects of what we now call social sustainability in relation to spatial qualities have been addressed in the discipline, e.g. in Mumford's (1964) essay on 'urbanity' meaning all that was commendable about the social life of cities, such as a civilised collective urban life and personal self-fulfilment;[1] in earlier views of 'urban life' such as Park's (1915) 'Human behaviour in the urban environment' and Wirth's (1938) 'Urbanism as a way of life'; and later, as 'vitality' and 'successful neighbourhoods' in Jane Jacobs' seminal work in the 1960s, in particular relation to architecture (Jacobs, [1961] 1993), or more recently as 'liveability' (James, 2015).

Recent connections still seem to be dominated by relevant but mainly normative aspirations and desirable qualities of cities. These aspirations are not off mark, but they do need to be carefully anchored. First, when searching for the precise material conditions of social sustainability, we should avoid teleological assumptions about urban and architectural design as a means to 'design the social world', largely based on an old-fashioned spatial determinism bordering on social control. However, it seems safe to assume that social sustainability does not happen in the ether. It is neither fully contingent nor random; it neither depends exclusively on space nor emerges in any spatial situation. Similar to other social phenomena, social sustainability *finds* material conditions, and those are what we need to understand. These conditions have to do with people's social and spatial abilities, such as accessing places, joining social situations, and engaging in opportunities to interact and create social relationships. They have to do with conditions of public spaces and cities to support and express public life as a manifestation of the public sphere. In short, they have to do with how urban spaces can foster socialities and sociabilities.

Second, material ideas of social sustainability would benefit from empirical evaluation, and therefore from approaches that can bring those connections and qualities to the forefront. Indeed, a key shortcoming in the usual treatments of social sustainability lies in the need for theoretical and empirical frameworks able to support rigorous empirical investigation (see Shirazi and Keivani, 2017).

This chapter attempts to do precisely these two things: it assesses certain material conditions of social sustainability and develops an approach that can handle its connections with the built environment through a rigorous method. However, objectifying such conditions and connections can prove exceedingly hard. For simplicity, it would be advisable to start with *elementary* factors in the relationship between the built environment and social sustainability, which could be grasped empirically in a relatively unproblematic fashion. We propose to do so by addressing how the built form can help (or hinder) a key condition of social interaction. Social interaction has been seen as an *indicator* of social sustainability in a number of ways. It relates to community participation, livelihood, and liveability, along with the construction of social ties, relationships, cohesion, and networks with effects on social capital (e.g. Bramley and Power, 2009; Dempsey *et al.*, 2012; Putnam, 2000). For instance, apparent causal connections were found between spatial factors like density, interaction opportunities, and the structure of social networks in a few selected areas (Raman, 2010).

In our turn, we would like to deal with a tangible material condition for the viability of social interaction: the *co-presence* of people in public space. Goffman (1972) addresses the conditions of co-presence whenever actors sense that they are close enough to be perceived and sense they are being perceived. We understand co-presence as bodies positioned within a spatial field where they can be perceived through sight or other senses, including rooms and public spaces. The awareness of others, in turn, may help build knowledge about different social groups, reduce anxiety associated with the contact with diverse people, and increase tolerance and empathy (Pettigrew and Tropp, 2008), which are important elements for the emergence of social sustainability through time. Importantly, co-presence is the bodily condition for the transition from the *perception* of others to face-to-face *interaction*, i.e. engaging in communicative exchange by discursive and non-discursive means. Going a step further, co-presence is 'fundamental to even the most elaborate forms of societal organization', a key feature in social integration (Giddens, 1984, p. 64). Methodologically, co-presence matters because actual interaction cannot be taken for granted in unplanned situations. It is too contingent. But we can be certain that interaction can emerge only if its material condition is met: people must be co-present (Netto, 2017). Finally, co-presence is an unambiguous phenomenon, and a basic component in the emergence of complex properties such as social connectedness, cohesion, capital, and so on.

We wish to relate the conditions of co-presence to detailed considerations of urban form, namely its architectural components. Of course we do not usually think about any role of architecture in social life, let alone in social sustainability. We take it for granted. We think *with* space all the time, but rarely *about* space (Hillier and Hanson, 1984) and how it matters in our daily lives. But we are more likely to do so when we notice *problems* in our neighbourhoods and communities. That is how our approach begins. Such problems might begin with that fact that urban form, despite being durable, is susceptible to change. Buildings can be replaced. New morphologies and architectural types can emerge and quickly transform neighbourhoods shaped for centuries in order to support social life. Indeed, entirely new urban areas may emerge. These processes can be found with particular intensity in regions of the world subject to extraordinary forces of urbanisation, like developing countries.

This chapter is particularly concerned with one of these challenging contexts: the erosion of the traditional architectural fabric in Brazil, where new architectural forms are shaping the urban landscape. These forms have been increasingly built as *isolated* buildings and gated communities, and resulted in rather *rarefied* urban patterns. We have witnessed an erosion of the architectural fabric in larger cities in Brazil coincidental with a progressive erosion of the fabric of co-presence in their public spaces. Understanding this process is a major concern of this chapter, and we hope to do so by looking into the elementary relations of built form and co-presence as an indicator of social sustainability. We approach such relations through the following steps:

- Examining the erosion of the urban fabric as a threat to social sustainability;
- Analysing empirically recognisable aspects of 'social sustainability', namely co-presence in streets, as a 'social effect of architecture';
- Introducing a method able to recognise objectively such effects;
- Applying this method on large-scale empirical studies in 46 randomly selected areas in 3 Brazilian capitals;
- Bringing our conclusions about the place of architecture in social sustainability, along with recommendations for future work and policy making.

The erosion of the urban fabric as a threat to social sustainability

Founded by the Portuguese in 1565, Rio de Janeiro is one the oldest cities in Brazil, and the capital of Brazil from 1815 to 1961, when Brasilia was built. It has around 7.5 million inhabitants and is the second largest city of the country. Colonised at the start of the eighteenth century by the Portuguese after the Madrid Treat was signed between Spain and Portugal, Porto Alegre is the capital of the southernmost state of Brazil, close to Argentina and Uruguay. It now has a population of around 1.5 million inhabitants. Florianópolis is older than Porto Alegre; it was founded by European immigrants in the seventeenth century and splits its 480,000 inhabitants between the island of Santa Catarina and the mainland.

The three cities share some idiosyncrasies: they have evolved over hilly topographies close to large bodies of water: Rio, by the Guanabara Bay and the Atlantic; Florianópolis, an island; and Porto Alegre, growing by Guaiba, a lagoon-shaped river. These peculiar geographical conditions shaped urban expansion radiating away from the initial settlement close to the waterfront (Rio and Porto Alegre), or into a handful of isolated fishing villages that were eventually connected by land through meandering roads that circumvent hills, mangroves, and lagoons in the island of Santa Catarina (Florianópolis).

Like others in Brazil, these cities faced extraordinary growth in the twentieth century, especially in the 1950s and 1960s. The threshold crossed in 2010, with more than 50% of the world population living in cities, was crossed in Brazil in 1965. In 2017, 76% of its 207.7 million population (i.e. 157.8 million people) lived in dense urban areas (Instituto Brasileiro de Geografia e Estatística [Brazilian Institute of Geography and Statistics] or IBGE, 2017). Until that time, Brazilian cities had mostly grown following the classic centre–periphery model, displaying a clear hierarchical structure formed by an old, dense, and somewhat vertical core with some radial roads with mixed land uses leading to residential interstices and sparse fringes. From the 1950s on, the growing use of private vehicles stimulated the sprawling of cities, allowing residents to live far from the central business district (CBD). By the end of the 1980s, global economic changes and the restructuring of the capitalist production system were felt in state regulations over Brazilian cities. Market-friendly urban policies spread, building height limits were loosened, and the construction of high-rise buildings in low-density residential areas was allowed. Today, Rio, Florianopolis, and Porto Alegre are cities that sprawl over their territories in geographically discontinuous ways, precariously connected by sparse street networks. They became unbalanced, polycentric systems where the residential fabric poorly fills the space between unevenly dispersed, high-density clusters.

Multi-storey buildings play a role in that sense. In Rio de Janeiro, they account for 37.62% of home types (with 54% for houses and 6.78% for residences in a community/condominium). The proportion in southern Brazilian capitals Florianópolis and Porto Alegre is 37.77% and 46.66%, respectively.[2] These numbers would go up if we included multi-storey architecture for commercial activities. Of course, these buildings are there for a reason. If the urban economics principle of a relationship between location and density is correct – and everything suggests that it is[3] – then multi-family housing responds to economic forces that aim to maximise profit and utility through a complex interplay of accessibility, land value, locational preferences, and purchasing power by multiple agents. Architecture, in a way, expresses such forces.

We mentioned that the real estate standards have recently changed in Brazil. Decades after the first modernist wave of dissolution of urban blocks (see Panerai *et al.*, 2004), a new wave has been brought to small-scale plots. The traditionally favoured type of multi-storey building, usually adjacent to other buildings in the block and adjacent to public spaces, was replaced by an *isolated* type of architecture, marked by disconnection from neighbouring buildings, large setbacks from the street, parking lots, walls, and railings. It has become the dominant feature in city making in Brazil, mainly due to increasing concerns about safety, incentives in zoning and building codes, and bigger profits achieved by higher floor area ratios. What is more, enclaves became associated with high-income groups, and are used as a way of expressing social differences and hierarchy in space (Caldeira, 2011). Decades of replication have left their mark on the urban landscape: fragmentation and vertical growth, a decline in the use of public space, and withdrawal of micro-economic activity to shopping centres and malls.

Urban growth shaped by high-rise buildings has led to the replacement of traditional morphologies, and to a discontinuous urban fabric. This morphological trend has coincided with decreasing levels of social appropriation, understood by many as a crisis of public spaces in Brazil (e.g. Figueiredo, 2012; Queiroga, 2017). At the same time, the reproduction of a *standard architectural* solution geared to particular niche markets, family budgets, and social classes seems increasingly detrimental to social diversity. In the long run, such trends could well lead to more spatially and socially segregated areas. These observations suggest a *background hypothesis* for the relationship between architecture and local conditions of social sustainability (Figure 7.1) related to the co-presence of people in streets.

In other words, if the dissolution of the social use of public space is related to the rarefication of urban form, then architecture has at least partly to answer for it. Factors of form, surface, and content of architecture would be active in their interface with public spaces and the social life that arises there. There is also the problem of the *activities* housed by architecture. They are fundamental in the use of public space, because they allow and help materialise relations in a local micro-economy. Would these activities emerge under any spatial condition? Urban economics since Alonso (1964) has shown that activities depend on certain conditions of urban density or demand, and that density is related to the economic diversity of urban areas. But the

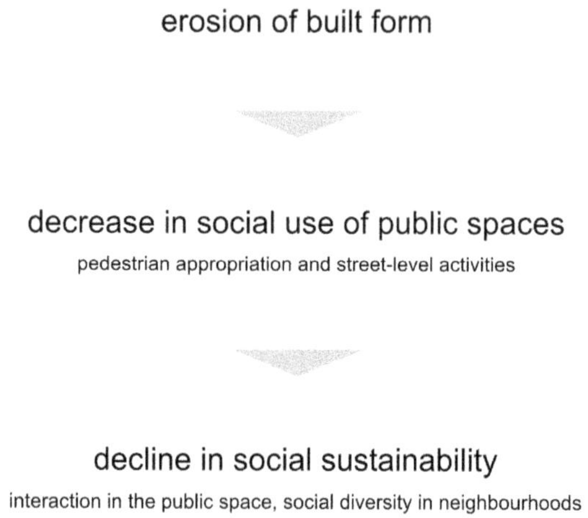

Figure 7.1 Background hypothesis: our research will cover empirically causal implications between the first two levels, while discussing connections with the third one.

materialisation of this relationship passes through architecture, and deserves attention, because they might be different for different typological arrangements.

The social effects of architecture

How could effects of spatial form occur? We need a refined hypothesis about what exactly in architectural form could have any role in stimulating life in public space. But, first, the great diversity of architectural forms needs to be analysed. We can do so by identifying and reducing such diversity into recurrent lexicons, as a result of characteristics that relate to certain buildings more than others. These lexicons generate *typologies*, a classic resource in architectural and urban theories, from historical typologies formulated by Caniggia and Maffei (2001) to analytical studies by Martin and March (1972) and Berghauser Pont and Marcus (2015). Types can be organised in various ways, according to substantive aims or cultural variations. In our case, elementary arrangements can be based on the relationship between built forms and their placement on the site as an unambiguous criterion for classification: namely, buildings with limits that coincide with the divisions of the urban plot – or, simply, 'continuous type' – or buildings free of connections with other buildings, termed here 'isolated'. Clearly, arrangements of these types can generate a virtually infinite variety of blocks and neighbourhoods, but we also have two archetypes: the *compact* block, showing built forms in direct association by their close adjacency, and the *fragmented* block, with free-standing buildings surrounded by large empty areas separated from the street by walls (Figure 7.2).

Figure 7.2 Architectural types as elementary arrangements of built forms in Rio's central business district (above) and Barra da Tijuca (below).

Source: Google Street view.

This typology is based on general geometric attributes, and their relationship with more detailed architectural features, such as density of doors or frontal setbacks, might be subject to variation. We can test whether this is an empirically coherent definition via statistics. We performed a *cluster analysis* on a database containing the architectural attributes of 4,038 buildings in one of our study cases, Porto Alegre, to check whether certain architectural features are particularly associated with a specific building type. Cluster analysis is a technique for grouping variables more frequently associated within a distribution, so that objects in a same cluster are more likely to appear together than those grouped in other clusters (Hair *et al.*, 1995). We used the correlation coefficient as the measure of statistical distance between objects, and applied the 'complete linkage' ('furthest neighbour') method, which is based on the evaluation of the maximum distance between two objects of different groupings.

This is a hierarchical procedure represented in a tree diagram or a *dendrogram*, like the one in Figure 7.3. Each variable progressively connects to a group of its own, and the nearest groups are agglomerated to form new groups in a next step of similarity. Figure 7.3 shows two groups with approximately a 40% degree of similarity: a first cluster aggregating *facade continuity*, *windows*, *doors*, and *open plots*, converging to the definition of the continuous type; and a second cluster aggregating *frontal* and *lateral setbacks*, *walls*, and

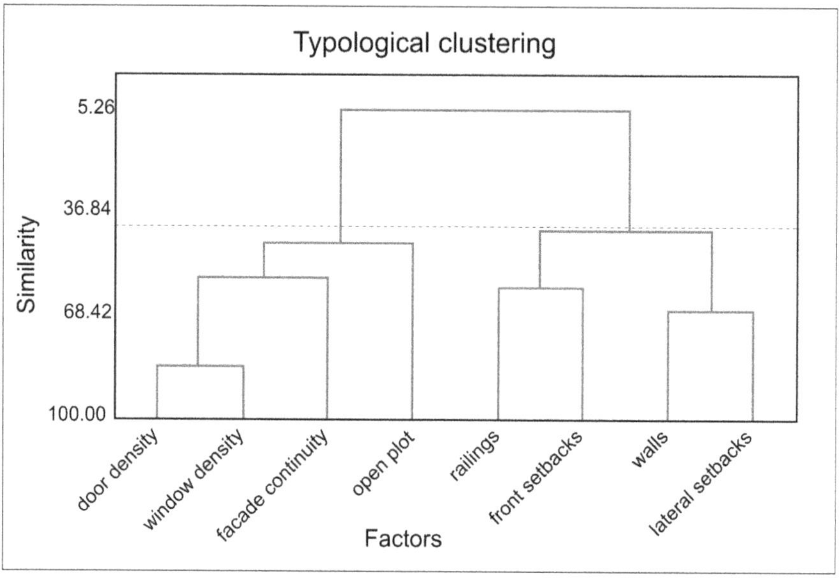

Figure 7.3 Dendrogram of architectural variables.

railings, converging to the isolated type. This analysis supports the conceptual definition of architectural types as a combination of a small set of architectural features, which could be visually recognised by trained researchers in the field, and measured in corresponding maps and plans.

The next step is to find out whether these basic architectural arrangements are associated with co-presence in the public space and with local micro-economic life in neighbourhoods. We know that bodily interactions depend on distance (see Allen 1977), which ultimately have to be travelled. If this material condition remains valid across scales, it will mean that the *continuous type* enables proximity and the *isolated type* tends to increase distances, adding more friction to movement. Therefore, our main hypothesis at this stage is that these *different architectural arrangements would have different effects on our presence in space,* with potential large-scale consequences. Other things being equal, the continuous type will respond more suitably to social life at a local level, by means of its contiguous façades. The isolated type would have the opposite effects, raising difficulties for pedestrian movement and activities, varying according to how far the buildings are set back from each other.

These spatial relationships of distance and proximity appear at the core of what we call 'the social effects of architecture' (Netto *et al.*, 2012; Saboya *et al.*, 2015; Vargas, 2017). Our aim is to identify such effects in different contexts, so spatial and social differences between cities and their inner areas are a fundamental part of the problem. These differences include symbolic and affective components, along with historical and geographical conditions that might influence the presence of people in public space. Recognising the significance of these forces, we need to see whether the effects of architecture on co-presence can be felt *even when all these differences and forces are at play*.

A method for grasping the social effects of architecture

How can the existence of these effects be tested, particularly when the actions of actors in the city are not just subject to architecture but also to factors such as the distribution of activities and centralities, and to differences in accessibility and mobility? Among the countless factors active in a socio-spatial system like the city, accessibility is generally suggested (in spatial economics and urban configuration studies) as one of the most important, because it incorporates the material aspect of extension and distance. If we want to identify clearly the existence and extent of the social effects of architecture, accessibility has to be controlled or, at least, monitored, because there is sound evidence in the literature that it is strongly related to pedestrian activity on the streets (e.g. Hillier *et al.*, 1993).

Another problem to consider is the possibility of accessibility having stronger effects as it increases, rendering the effects of architectural morphology less perceptible. Our method assesses accessibility at different levels and, once it has been methodologically controlled, we can approach the architectural characteristics of a large number of areas and compare them statistically with social factors. Then we can verify whether such coincidences have material and probabilistic significance.

Of course the social effects of architecture might be felt in different spaces, such as suburbs and rural areas, but we focused on urban areas with the (non-exclusive) presence of multi-family typologies as a way of avoiding both suburban morphologies and informal settlements.[4] We assessed accessibility via configurational measures, a kind of network analysis (see Hillier *et al.*, 1993) that considers the street segment as the unity of study, as opposed to accessibility models based on aggregate zones. We used a measure of *closeness centrality* (also known as 'global integration') in Rio e Porto Alegre and *betweenness centrality* (or 'choice') in Florianopolis. The former calculates topological distances as changes of direction within the network and expresses the ease of reaching each space, and the latter considers how often a given segment lies on the shortest routes between all possible pairs of segments. We broke each city's accessibility range of values into 20 levels, from which three were selected – low, middle, and high.

We also analysed population density in three levels derived from the census tracts distribution. Our sample was randomly selected within these subsets of streets and census tracts as spatial units (Table 7.1). In order to have diversity in our sample, we adopted an experimental design that combined three levels of population density and high, medium, and low levels of configurational accessibility, resulting in nine combinations in which street segments were selected and surveyed.

Table 7.1 Cities and number of areas, street segments, and buildings sampled

	Rio de Janeiro	*Florianópolis*	*Porto Alegre*
Areas	24	19	3
Street segments	250	174	298
Buildings	4,174	1,225	4,038

Source: authors.

We explored a variation of this method in Porto Alegre. Keeping the same premises of the experimental design, we aimed at local variations in accessibility while bringing into play a component hitherto treated as noise: the transit stations, a factor capable of deforming pedestrian movement patterns. We selected three large areas (550 metres in radius) around transit stations sharing the same high level of *global* accessibility, and stratified the street segments as low, medium, and high levels considering their *local* accessibility values (Figure 7.4).

The distribution of activities and architectural features was measured to comprise a total of approximately 10 social and 40 spatial variables such as the number of windows and doors, the presence of walls or fences, front and lateral setbacks, and so on. These variables were largely derived from urban design literature and correspond to those more frequently associated with compact and walkable neighbourhoods (Gehl, 2011; Ewing and Clemente, 2013; Jacobs, [1961] 1993; Moundon, 1992). Local activities were considered in basic categories such as residential, institutional, and commercial land uses (retail, service, or both), which were counted at street level and on upper floors. We assessed diversity of activities through Shannon's (1948) entropy and the Gini–Simpson index (Jost, 2006), considering the number of different categories and how evenly they were distributed. The diversity index increases when the number of different activities and their evenness increases in a given segment. Street segments in which categories are present in equal shares have the highest diversity index.

Our aim here was to compare the distribution of architectural features and ground-level land uses with pedestrian activity, and see whether consistent correspondences could be found. In order to quantify pedestrian activity, systematic observations of moving and stationary pedestrians were carried out at each sampled street segment in six periods (from 2.5 minutes to 5 minutes each) during typical workdays, from 9 a.m. to 7 p.m.

Empirical findings

Pedestrians and building types

Streets where the continuous type is predominant tend, on average, to have more than twice as many moving pedestrians as in the isolated type, for all

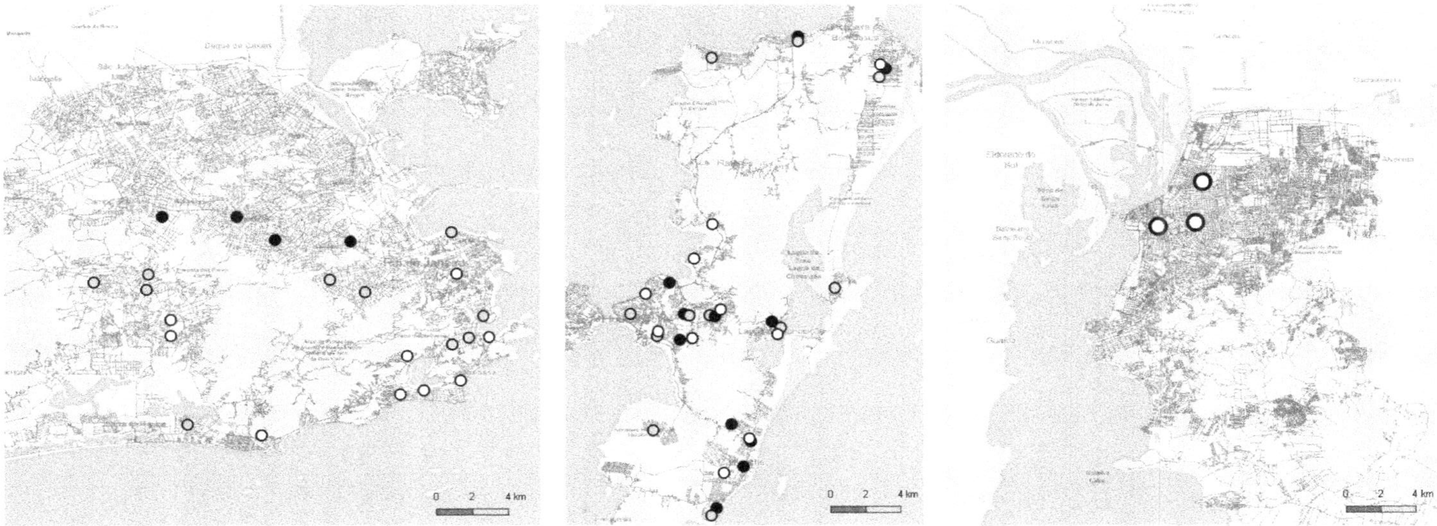

Figure 7.4 From left to right: selected areas in Rio de Janeiro, Florianópolis, and Porto Alegre. Accessibility levels are represented by white (low), grey (medium), and black (high) dots; for Porto Alegre, dots represent areas with similar global accessibility.

Source: data from Mapbox and municipal records.

street segments analysed. In Rio's low-accessibility areas, moving pedestrian averages are around three times higher in street segments with more than 50% of continuous types than in those with less than 50% ($P < 0.01$).[5] In high-accessibility areas this trend is weaker, but the pedestrian activity is still larger on streets with a predominance of continuous types. Our box plot graphs (Figure 7.5) bring this and other results.

Pedestrians and façade continuity

Our hypothesis predicts a relationship between façade continuity along blocks and what happens in the public space in front of it. Continuity was measured as the ratio between the sum of façades' width and the total sum of front property lines' lengths. When the façade-continuity indices are broken into intervals, we can see a trend of increasing pedestrian movement, reaching a steep curve in continuity indexes above 85% in Florianópolis, and 95% in Rio and Porto Alegre. The greater the continuity, the greater the levels of pedestrian movement.

Pedestrians and setbacks

Analysis of average *distances between buildings* in blocks (street segment) in Rio shows that pedestrian movement decreases as those distances increase, from around 14 pedestrians where average distances are less than 2.5 m, to about 3 pedestrians for distances between 15 and 20 m. We also observed a similar reduction in pedestrian movement for increasing distances between buildings and the street, the *frontal setback*: from 11.5 moving pedestrians per minute where average distances are less than 1 m, to 2.3 pedestrians for distances greater than 5 m. Porto Alegre and Florianópolis show remarkable similarities.

Pedestrians and windows

This relationship recalls Jacobs's ([1961] 1993) proposition of a positive relationship between the number of windows and life in the streets. Corroborating the Jacobsian hypothesis, we found a clear pattern: the increase in the *density of windows* (number of windows per metre of street segment) is followed by an increase in the number of pedestrians in the streets. A similar relationship was found between pedestrians and *door density*.

Pedestrians and open plots and walls

Something similar occurs when we consider the absence of *barriers* between buildings and the street: the higher the presence of direct interfaces, free from walls and railings, the higher the number of moving pedestrians in the streets. In Rio, pedestrian numbers range from an average of 0.8 pedestrians per minute where there are less than 0–5% direct building–street interfaces to an

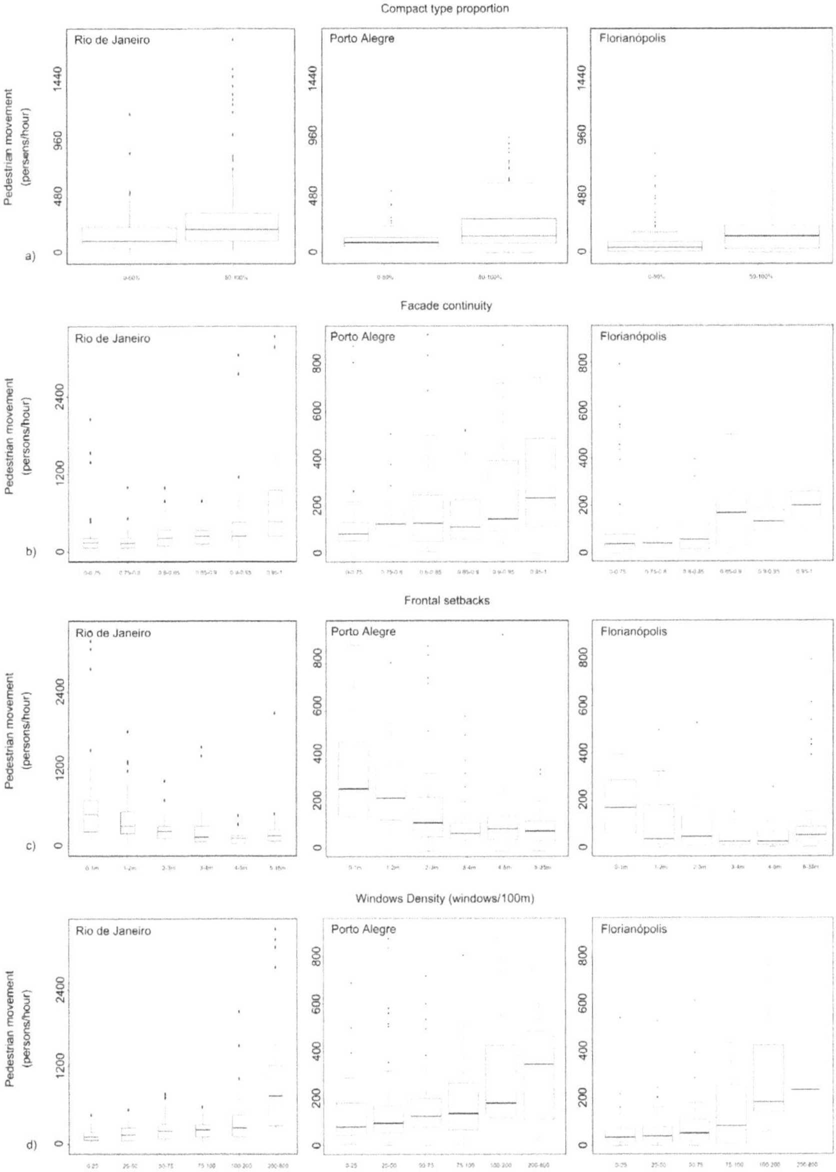

Figure 7.5 Moving pedestrian averages in relation to architectural features in Rio de Janeiro, Florianópolis and Porto Alegre (all accessibility levels), from top to bottom: (a) in street segments with low (<50%) and high (>50%) proportion of continuous types; (b) facade continuity; (c) frontal setbacks; and (d) windows density.

average of 5.3 pedestrians per minute, when 70% or more of the plots are open to the street. Numbers fall systematically when walls become dominant in the building–street interface, dropping to an average of 1.8 pedestrians per minute moving along street segments, with 70% or more of plots displaying front walls.

Exploring statistical correlations,[6] we also identified trends consistent with our hypothesis, but more so in areas of low accessibility. Positive correlations were found between stationary and moving pedestrians and the continuous type, whereas correlations with the isolated type are almost diametrically opposed in Rio and Florianópolis (Tables 7.2 and 7.3). We captured similar trends through other key morphological variables such as the *continuity index*, *side and front setbacks*, and *building interfaces* with public spaces.

Overall, the combination of correlations supports the notion that the continuous type favours *a direct relationship* between architecture and public space, and that this has a positive association with the presence of pedestrians and activities. Correlations for high-accessibility areas are weaker; yet they follow

Table 7.2 Pearson's correlations for spatial and social factors in Rio and Florianópolis: low-accessibility areas

Spatial and social factors	*Stationary pedestrians*		*Moving pedestrians*	
	Rio de Janeiro	*Florianópolis*	*Rio de Janeiro*	*Florianópolis*
Continuous type	0.41**	0.58**	0.33**	0.58**
Isolated type	–0.41**	–0.51**	–0.34**	–0.66**
Continuity index	0.46**	0.29**	0.42**	*0.41*
Frontal setback	–0.39**	–0.23*	–0.42**	*–0.31*
Open plot	0.55**	0.31	0.63**	*0.30*
Windows	0.51**	0.48**	0.73**	0.62**

Notes
P: <0.1; * <0.05; ** <0.01.

Table 7.3 Pearson's correlations for spatial and social factors in Rio and Florianópolis: high-accessibility areas

Spatial and social factors	*Stationary pedestrians*		*Moving pedestrians*	
	Rio de Janeiro	*Florianópolis*	*Rio de Janeiro*	*Florianópolis*
Continuous type	0.21	0.32*	0.07	0.07
Isolated type	–0.22	–0.47**	–0.04	–0.21
Continuity index	0.42*	0.25	0.12	*0.25*
Frontal setback	–0.25	–0.16	–0.11	–0.19
Open plot	0.46**	0.43**	0.41**	0.46**
Windows	0.18	0.44**	*0.25*	0.59**

Notes
P: <0.1; * <0.05; ** <0.01.

the trends of positive and negative signs found in low-accessibility areas in these two cities.

In turn, our study in Porto Alegre focused on finely grained spatialities captured by *local accessibility* measured in buffers of 550 metres around three chosen transit stations, all of them located within the city's global accessibility core (Table 7.4). Each selected area encompasses a main street where transit stations are located, some adjacent and near street segments (i.e. at one or two topological steps), and eventually sets of streets relatively distant from the main street.

Correlations between spatial features and pedestrian activity are in general statistically significant and follow closely those found in Rio and Florianópolis – especially in high local accessibility streets. On low local accessibility street segments, correlations fall sharply. We see strong correlations for moving pedestrians, the *continuous type*, and *open plots*.

Architecture and street-level economic activities

A micro-economy can fully express itself locally only if certain architectural conditions are present. Buildings must be able to 'engage' with people or generate *affordances* (Gibson, 1979) for pedestrians appropriating public spaces while searching for activities, and offer spatial conditions for activities to emerge and attract pedestrians (see Carmona, 2015). We found negative correlations between the isolated type and non-residential activities such as retail and services. We also found a decline in activity diversity when the isolated type becomes predominant. These trends are more expressive in low-accessibility areas (Table 7.5). Other items associated with the isolated type, such as *walls* and *frontal setback*, are less likely to couple with commercial activities.

This does not mean that the local economy became 'smaller', because the percentage of urban plots with shops tends naturally to be relatively small in

Table 7.4 Pearson's correlations for spatial and social factors in Porto Alegre (local accessibility)

Spatial and social factors	*Stationary pedestrians*		*Moving pedestrians*	
	Low local accessibility	*High local accessibility*	*Low local accessibility*	*High local accessibility*
Continuous type	–0.02	0.23*	0.09	0.39**
Isolated type	–0.01	*–0.21*	*–0.17*	–0.40**
Continuity index	0.02	0.02	0.22*	0.37**
Frontal setback	0.14	–0.04	–0.20*	–0.41**
Open plot	–0.29	0.09	0.41*	0.31
Windows	0.10	0.12	0.31*	0.41**

Notes
P: <0.1; * <0.05; ** <0.01.

Table 7.5 Local economic activities and architectural factors in different levels of global accessibility in Rio de Janeiro e Florianópolis, and local accessibility in Porto Alegre: Pearson's correlations

Spatial and social factors	*Commercial activities and services Low-accessibility areas*			*Commercial activities and services High-accessibility areas*		
	Rio	*Florianópolis*	*Porto Alegre*	*Rio*	*Florianópolis*	*Porto Alegre*
Continuous type	0.42★★	0.48★★	0.24★★	0.33★	0.17	0.35★★
Isolated type	–0.45★★	–0.48★★	–0.26★★	–0.34★	–0.23	–0.31★★
Continuity index	0.38★★	0.40★★	0.22★★	0.39★	0.27★	0.34★★
Frontal setback	–0.29★★	–0.13	–0.14	–0.18	–0.21	–0.42★
Windows	0.52★★	0.26★	0.31★★	–0.19	0.10	0.42★
Open plot	0.65★★	0.19	0.08	0.53★★	0.54★★	0.21

Notes
P: <0.1; ★ <0.05; ★★ <0.01.

any city (in Rio only 7.56% of plots involve activities other than housing, teaching, and healthcare) but that it probably shifted from the traditional street shops to places like shopping centres. The main problem here is the breaking of the *convergence* between accessibility patterns and location of commerce due to the growing election of an architecturally isolated type antithetical to street commerce, which is exactly what can be seen in major Brazilian cities. Finally, the high percentage of residential streets could also stimulate more pedestrians if any potential for densification in multi-storey buildings did not involve the isolated type, usually coupled with walls or fences and setbacks.

Of course, correlations and other basic statistical techniques are mere indications of *association* rather than *causality*. Nevertheless, there are ways in which these factors can be investigated more conclusively than through simple bivariate analyses. In order to identify the joint contribution of architectural features to pedestrian activity, we explored multiple linear regressions using the ordinary least squares (OLS) technique. We approached pedestrian movement as a probabilistic function of *accessibility*, *density*, *activity*, and *architectural features*. Once global accessibility had been controlled in our first procedure, this final analysis showed that a small number of spatial factors accounts for a substantial portion of the distribution of pedestrian movement in Rio (Table 7.6).

Showing a highly adjusted coefficient of determination ($R^2 = 0.94$) with statistical significance at the 99% confidence level, the result for the aggregate sample (all areas in Rio) corroborates the positive relationship between the architectural factors associated with the continuous type and pedestrian movement. We can understand the coefficients as the rate of change in the number of pedestrians per minute per one-unit change of the predictors (local accessibility, dwelling density, and so on), assuming all others are held constant. Therefore, the coefficient of 0.39 for dwelling density indicates an increase of 0.39 pedestrians per minute when we add one dwelling unit per metre in a street segment.

Table 7.6 Multiple linear regression: explanatory potential of spatial variables on pedestrian movement

Predictors	*Coefficient*	*P*	R^2	*Shapiro–Wilk test*★
Local accessibility	0.50	<0.001	0.94	0.36
Dwelling units' density	0.39	<0.001		
Diversity (ground-floor level)	1.67	<0.001		
Continuous type	0.33	0.03		

Note

★ Shapiro–Wilk test: this test is used to assess whether the hypothesis of data normality/errors is correct. If the *P* value is greater than the level of significance (in this case 0.05), we do not reject the null hypothesis that data and errors have a normal distribution. In this case, the *P* value is 0.36, indicating that the assumption that errors have a normal distribution in the linear regression model is not being violated.

Finally, we used the partial least squares (PLS) form of regression, which models the relationship between predictor variables and more than one response at the same time. This is a technique suitable for datasets where many variables show high correlations between themselves ('co-linearity'). The PLS regression also groups the independent variables in components or factors according to proximity, just like the cluster analysis. Figure 7.6 shows the intensity of each predictor's impact over moving and stationary pedestrians, based on the spread of lines representing each factor: positive ones to the right, negative to the left.

In short, different analyses brought robust statistical evidence of the association between architectural features and social outcomes:

- Open plots, window density, façade continuity, and the continuous building type are *positively* associated with pedestrian activity;
- The same architectural features have a clear *positive* relationship with the distribution of economic activities and land use diversity on the streets;
- Walls, lateral and frontal setbacks, and the isolated building type have *negative* associations with pedestrian activity;
- The same architectural features have a clear *negative* relationship with the distribution of economic activities on the streets.

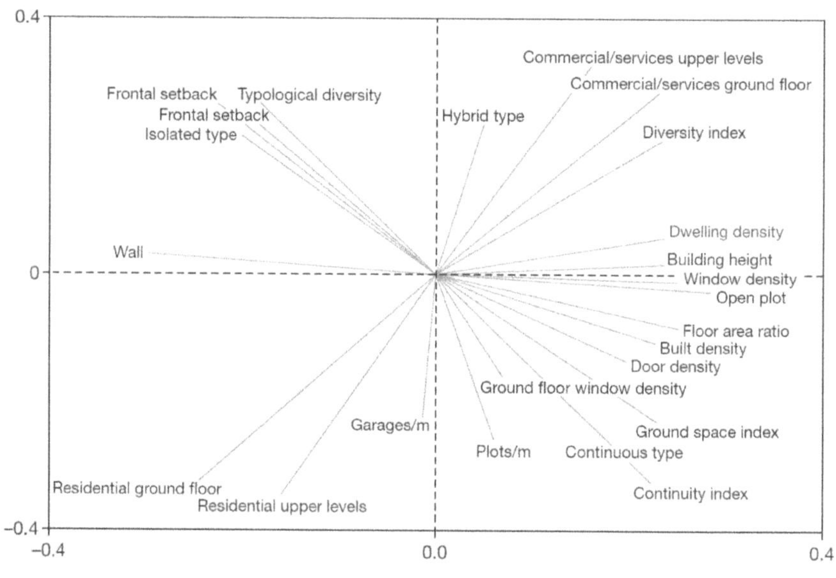

Figure 7.6 PLS regression analysis: the right-hand spatial factors show positive behaviour in relation to pedestrian movement. The length of lines for each factor shows their level of importance.

Concluding remarks: the place of architecture in social sustainability

Our main idea in this chapter was to find whether, and to what extent, architecture matters for basic factors of social sustainability, namely the presence of people in public space as a condition and local activities as motivators of social interaction. The fact that *we could find consistent relationships within complexities and contingencies in the connections of the 'social city' to the 'physical city' is surprising*. The spatial disciplines (urban studies, spatial economics, human geography) offer plenty of evidence that space matters, but our research finds that architectural features close to the bodily appropriation of space are an active part of this relationship. Our results indicate that the new architectural type most present in Brazilian cities has a significant negative association with elementary factors of social sustainability (Figure 7.7).

Findings in three Brazilian state capitals show considerable convergences, particularly in low-accessibility areas, and interesting intensity differences concerning the role of high accessibility in relation to the role of architecture. But they also show that the power of *contextual differences* over general spatial properties can be neither ignored nor assumed, as is often the case. Three cities clearly do not offer the force of generalisation, but they do allow the appreciation that the *universality* of architectural influences, if confirmed, is subject to contextual variations. On the other hand, *contextual variations do not appear to render* co-presence a contingent fact, as relativist positions geared to ignore causal properties of urban space would have. The weight of contextual differences therefore also requires empirical demonstration, along with the challenge of understanding what the active causalities and unpredictabilities at play are.

This empirical approach should now look more closely into contextual differences and idiosyncrasies. A major advantage is its reproducibility, which is useful to assess levels of social appropriation of public spaces and activity diversity in the built environment in other regions, cities, and urban areas. Strong findings may lead to predictive analysis of alternative scenarios, and we are now developing a computational model that can simulate such scenarios from parameters empirically identified.

What can be said about the architecture that dominates urban production in Brazil and elsewhere, with its negative influence on the pedestrian? The predominant architectural type today breaks the relationship between built

ISOLATED TYPE	FEATURES	NEGATIVE EFFECTS OVER
	low façade continuity	pedestrian movement
	large frontal setbacks	stationary people in public space
	walls and railings	commercial activities
	low window and door densities	activity diversity

Figure 7.7 Associations between the isolated type and local social factors.

form and open forms of social life. Isolated types have been increasingly positioned on the more accessible streets, ignoring their pedestrian potential, which ultimately reduces the natural attraction of these streets. This reduction imposes a *divergence* of urban patterns, breaking with the mutual correspondence between densities, accessibility, activity diversity, and movement (Netto *et al.*, 2012).

Standardisation of architectural models focused on specific market niches and social classes, and social segregation, is a trend that can be seen clearly but that is generally ignored precisely because these models are so pervasive. The powerful self-referential logic of production of standard architectural solutions ultimately shapes the practice of architects unaware of its negative implications. A fragmented landscape of isolated objects reflects this splintered logic, centred on the individuality of decisions and unconcerned with systemic consequences. It is all reflected in the Brazilian urban landscape. The unintended consequences of choosing an architectural type tend to be revealed only later, when choices have been fixed and we resign ourselves to the idea that they are here to stay.

Invisible effects influencing the overall path of a city as a fabric of interactions may remain beneath the resolution of the planners' lenses. Indeed, planners face a permanent challenge when designing building codes: simultaneously to consider the urban built stock as a private asset with great economic value and as a part of the urban system, where every built form has effects throughout the social and spatial fabric. In developed countries, performance-evaluation planning is now well established, operating through parameters of architectural form to minimise its impact on the environment. Even though there are 'smart growth' guidelines and form-based codes that seek to induce the compact and diverse city, few instruments seem to be directly aimed at strengthening communities and promoting sociocultural diversity.

In Brazil, master plans tend to be more rudimentary: they are basic sets of building regulations centred on the individual land plot with little attention paid to the systemic performance of new buildings in relation to the built environment. They seek to maximise building potential based on the vague idea that high densities reduce urbanisation costs and make cities more sustainable. In turn, strategies to accomplish the right to the city were epitomised in the well-known 'Statute of the City' bill (Brazilian Government, 2001), a federal law that introduced innovative mechanisms in urban planning especially focused on issues of social inclusion. After a period of genuine social empowerment that promoted a wave of successful participatory planning spanning two decades, however, civic participation did not resist the 2010s' real estate market offensive and communities were dismissed from the urban debate in the country.

Thus, the social effects of architecture and the impact of building choices on pedestrian activity still have a long way to go before they become a genuine component of participation arenas, despite the evidence gathered by studies such as ours. Do dwellers, shoppers, and local entrepreneurs prefer walls and gated communities as a means for privacy and alleged safety? Or would they trade it for greater sociocultural diversity, more economic vitality, and stronger social ties? This could be the subject of other research efforts, combining in-depth qualitative and quantitative methods, as well as ad-hoc discussions concerning specific settings and decision-making contexts. Either way, ill-informed choices will most likely result in undesired outcomes. These decisions should be shaped through empirically informed debates, needed especially in contexts of fragile institutional frameworks, so the trade-offs are sufficiently clear. What is clear, however, is that cities and communities interested in creating and sustaining lively, vibrant streets must not underestimate the importance of architectural form. The available evidence suggests that there is a significant association between these kinds of streets and compact, continuous, and permeable buildings, and that careful consideration must be given to short- as well as long-term consequences of decisions concerning the built environment.

Acknowledgements

The National Council of Research (CNPq) in Brazil supported this research. We would like to thank our team of research collaborators for their statistical and empirical support.

Notes

1 Mumford's essay can be found in his 1964 book *The Highway and the City*.
2 Of urban addresses in Rio de Janeiro 90.6% are residential; in Florianópolis the figure is 90.8%; in Porto Alegre 91.1% (IBGE, 2010; CNEFE – Cadastro Nacional de Endereços para Fins Estatísticos).
3 This principle was implied by Alonso's (1964) model of location. Our own research found evidence of it.
4 Due to striking idiosyncrasies in the way public space supports co-presence and interaction in these areas, seen in Aguiar (2010) and Marques (2012), we did not include informal settlements in the areas surveyed.
5 Methodologically, we performed a two-sample student's *t*-test to assess the significance of the differences between the average number of pedestrians in two distinct classes. The test of statistical significance (*P* value) examines the probability that the difference between the means arises by chance. *P* values that >0.05 have no statistical significance, according to the conventionally adopted parameter of 95% confidence.
6 Correlations based on Pearson's coefficient vary between –1 and +1 (perfect negative or positive correlation, with zero denoting no correlation).

References

Aguiar D (2010) *Alma Espacial – Corpo e o Movimento na Arquitetura* [*Spatial Soul – Body and movement in Architecture*]. Porto Alegre: Ed. UFRGS.

Allen T (1977) *Managing the Flow of Technology*. Cambridge MA: MIT Press.

Alonso W (1964) *Location and Land Use: Toward a General Theory of Land Rent.* Cambridge, MA: Harvard University Press.

Anand S and Sen A (1996) Sustainable human development: Concepts and priorities. Office of Development Studies Discussion Paper, No. 1, UNDP, New York.

Berghauser Pont M and Marcus L (2015) What can typology explain that configuration can not? In: *Proceedings of the 10th International Space Syntax Symposium.* London: University College London.

Bramley G and Power S (2009) Urban form and social sustainability: The role of density and housing type. *Environment and Planning B: Planning and Design* 36(1): 30–48.

Brazilian Government (2001) City Statute: Federal Law 10.257. Brasília: House of Representatives.

Caldeira T (2011) *Cidade de muros: crime, segregação e cidadania em São Paulo* [*City of Walls: Crime, Segregation, and Citizenship in São Paulo*], 3rd edn. São Paulo: Ed. 34.

Caniggia G and Maffei GL (2001) *Architectural Composition and Building Typology: Interpreting Basic Building.* Florence: Alinea.

Carmona M (2015) London's local high streets: The problems, potential and complexities of mixed street corridors. *Progress in Planning* 100: 1–84.

Dempsey N, Brown C, and Bramley G (2012) The key to sustainable urban development in UK cities? The influence of density on social sustainability. *Progress in Planning* 77(3): 89–141.

Ewing R and Clemente O (2013) *Measuring Urban Design: Metrics for Liveable Places.* Chicago, IL: Island Press.

Figueiredo L (2012) Desurbanismo: um manual rápido de destruição de cidades [Disurbanism: A quick guide to city destruction]. In: Aguiar D and Netto VM (eds), *Urbanidades.* Rio de Janeiro: Folio Digital.

Gehl J (2011) *Life between Buildings: Using Public Space.* Washington, DC: Island Press.

Gibson J (1979) *The Ecological Approach to Visual Perception.* Boston, MA: Houghton-Mifflin.

Giddens A (1984) *The Constitution of Society: Outline of the Theory of Structuration.* Cambridge: Polity Press.

Goffman E (1972) *Interaction Ritual.* London: Allen Lane.

Hair Jr JF, Anderson RE, Tatham RL, and Black WC (1995) *Multivariate Data Analysis*, 4th edn. Englewood Cliffs, NJ: Prentice Hall.

Hillier B and Hanson J (1984) *The Social Logic of Space.* Cambridge: Cambridge University Press.

Hillier B, Penn A, Hanson J, Grajewski T, and Xu J (1993) Natural movement: Or, configuration and attraction in urban pedestrian movement. *Environment and Planning B: Planning and Design* 20(1): 29–66.

Instituto Brasileiro de Geografia e Estatística (IBGE) [Brazilian Institute of Geography and Statistics] (2010) Cadastro Nacional de Endereços para Fins Estatísticos [National Address Cadastre for Statistical Purposes]. Available at: https://ww2.ibge.gov.br/home/estatistica/populacao/censo2010/cnefe/default_cnefe.shtm (accessed 15 November 2018).

Instituto Brasileiro de Geografia e Estatística (IBGE) [Brazilian Institute of Geography and Statistics] (2017) Classificação e caracterização dos espaços rurais e urbanos do Brasil: uma primeira aproximação [Classification and characterization of rural and urban spaces in Brazil: A first approximation]. Coordenação de Geografia, Rio de Janeiro.

Jacobs J (1961) 1993. *Life and Death of Great American Cities*. Reprint, New York: Modern Library Edition.
James P (2015) *Urban Sustainability in Theory and Practice*. London: Routledge.
Jost L (2006) Entropy and diversity. *Oikos* 113(2): 363–375.
McKenzie S (2004) *Social Sustainability: Towards Some Definitions*. Magill: Hawke Research Institute.
Marques EC (2012) Social networks, segregation and poverty in São Paulo. *International Journal of Urban and Regional Research* 36(6): 958–979.
Martin L and March L (eds) (1972) *Urban Space and Structures*. Cambridge: University Press.
Marshall G (ed.) (1998) *Dictionary of Sociology*. Oxford: Oxford University Press.
Moudon AV (1992) Getting to know the built landscape: Typomorphology. In: Franck K and Schneekloth L (eds), *Ordering Space. Types in Architecture and Design* (pp. 289–311). New York: Van Nostrand Reinhold.
Mumford L (1964) *The Highway and the City*. London: Secker & Warburg.
Netto VM (2017) *The Social Fabric of Cities*. London, New York: Routledge.
Netto VM, Saboya RT, Vargas JC, Figueiredo L, Freitas C, and Pinheiro M (2012) The convergence of patterns in the city: (Isolating) The effects of architectural morphology on movement and activity. In: *Proceedings of the 8th International Space Syntax Symposium*. Santiago, Universidad Católica de Chile.
Opp SM (2016) The forgotten pillar: A definition for the measurement of social sustainability in American cities. *Local Environment* 22(3): 286–305.
Panerai P, Castex J, DePaule J-P, and Samuels I (2004) *Urban Form: The Death and Life of the Urban Block*. New York: Routledge.
Park RE (1915) The city: Suggestions for the investigation of human behaviour in the urban environment. *American Journal of Sociology* 20(5): 577–612.
Pettigrew TF and Tropp LR (2008) How does intergroup contact reduce prejudice? Meta-analytic tests of three mediators. *European Journal of Social Psychology* 38(6): 922–934.
Putnam RD (2000) *Bowling Alone: The Collapse and Revival of American Community*. New York: Simon & Schuster.
Queiroga E (2017) Lugares públicos: atravessamentos entre espaços livres e edificados, públicos e privados [Public places: The overlapping of open and built, public and private spaces]. In: Netto VM, Saboya RT, Vargas JC, and Carvalho T (eds), *Efeitos da Arquitetura: Impactos da Urbanização Contemporânea no Brasil* [*Effects of Architecture: Impacts of Contemporary Urbanisation in Brazil*]. Brasília: FRBH Edições.
Raman S (2010) Designing a liveable compact city. Physical forms of city and social life in urban neighbourhoods. *Built Environment* 36 (1): 63–80.
Saboya RT, Netto VM, and Vargas JC (2015) Fatores morfológicos da vitalidade urbana: uma investigação sobre o tipo arquitetônico e seus efeitos [Morphological factors of urban vitality: An investigation on architectural types and its effects]. *Arquitextos* 180(2).
Sen A (2013) The ends and means of sustainability. *Journal of Human Development and Capabilities: A Multi-Disciplinary Journal for People-Centered Development* 14(1): 6–20.
Shannon CE (1948) A mathematical theory of communication. *Bell System Technical Journal* 27(3): 379–423.
Shirazi MR and Keivani R (2017) Critical reflections on the theory and practice of social sustainability in the built environment: a meta-analysis. *Local Environment*: 1526–1545.

Vargas JC (2017) Forma urbana e transporte a pé: mobilidade, caminhabilidade, vitalidade [Urban form and pedestrian movement: Mobility, walkability, vitality]. In: Netto VM, Saboya R, Vargas JC and Carvalho T (eds), *Efeitos da Arquitetura: Impactos da Urbanização Contemporânea no Brasil [Effects of Architecture: Impacts of Contemporary Urbanisation in Brazil]* (pp. 71–89). Brasília: FRBH Edições.
Wirth L (1938) Urbanism as a way of life. *The American Journal of Sociology* XLIV: 1.
Woodcraft S, Bacon N, Hackett T, and Caistor-Arendar L (2012) *Design for Social Sustainability: A Framework for Creating Thriving New Communities*. London: Social Life/Young Foundation.

8 Social sustainability in eco-urban neighbourhoods

Revisiting the Nordic model

Meg Holden, Annika Airas, and Majken Toftager Larsen

Introduction: the Nordic model and social sustainability

The so-called 'Nordic model' of governance is the envy of many advocates of sustainability, fairness, and well-being. In this chapter, we consider the Nordic model and some of the eco-urban neighbourhoods that this model has generated from past to present, in order to ask in what way this model does not serve social sustainability goals. More specifically, we interrogate the fit between the internal coherence of the integrated approach to social sustainability that has long been in evidence in the Nordic welfare state model and the approach to staging models of socially sustainable development in urban neighbourhoods that has emerged, more recently, internationally.

According to the OECD (Organisation for Economic Co-operation and Development) 'The inclusive Nordic welfare model, and its long tradition for social investment, has been discussed with fascination and often admiration for decades' (OECD, 2016). A welfare state model is classified as an approach to government in which

> the state plays a key role in the protection and promotion of the economic and social well-being of its citizens … based on principles of equality of opportunity, equitable distribution of wealth, and public responsibility for those unable to avail themselves of the minimal provisions for a good life.
>
> (Iqbal and Todi, 2015: 338)

The successes of Nordic countries are often recognised in the form of international rankings in measures from 'economic competitiveness to social health to happiness' (*The Economist*, 2013b). For example, Denmark, Finland, Norway, and Sweden rank within the top ten in league tables measuring such diverse conditions as the Social Progress Index (Social Progress Imperative, 2017), World Happiness Report (Helliwell *et al.*, 2017), Corruption Perception Index (Transparency International, 2017), and Income Inequality (OECD, 2014).

Popular depictions of the Nordic model's accomplishments include quality public health care, good education, generous parental leaves, economic competitiveness, modern technology, transparency, taxation, post-retirement and unemployment benefits, low corruption and gender equality, among others (see *The Economist*, 2013a, 2013b; Milne, 2014; Iqbal and Todi, 2015).

Moreover, Nordic nations have maintained such broad-based achievements as they have weathered economic crisis: serious crises have been faced successively in the 1980s and 1990s by different Nordic countries. Responding to these crises has meant significant changes to welfare model practices such as employment, housing, and family policy (see *The Economist*, 2013a, 2013b). The extent of these changes has led Andersen and Roine (2015: 7) to question the very notion of a Nordic model: 'Is it really meaningful to talk about the existence of such a thing?' Maybe the answer is 'no', in terms that can be ascribed to specific policies. At the same time, the Nordic nations persist in adhering to a common set of encompassing goals and principles:

> Goals – such as equal opportunity in life regardless of family background, the eradication of poverty, gender equality, the lowering of income inequality, etc. – as well as some principles – such as individually based universal rights to things such as health care and education, well-organized labour markets, etc. – have remained largely stable.
>
> (Andersen and Roine, 2015: 8–9)

Notably, the idea of 'social sustainability', while having a wide variety of definitions, shares many characteristics with this model. Even though sustainability discussions since the late 1980s have primarily emphasised environmental and/or economic perspectives, 'Social sustainability is an area of intense local political debate that draws from different perspectives about the purpose and best means of urban development and priorities for addressing human and community needs' (Holden, 2012: 239). Holden cites Partridge (2005) when pointing out that common social sustainability themes include, for example 'quality of life, equity, inclusion, access, a future focus and participatory process' (Holden, 2012: 530). Although social sustainability remains the least-studied aspect of sustainable development (Shirazi and Keivani, 2017), and sometimes is difficult to reconcile with the environmental and economic objectives of sustainability, urban policy makers find the notion of social sustainability increasingly difficult to ignore.

The image of the Nordic model integrates the environmental, economic and social aspects of the sustainable development ideal more seamlessly than most. For example, in *The Economist*, a correspondent summarised the 'spirit of the Nordic region' by evoking images of a Danish mother cycling, a Swedish father enjoying lunch with his small children, and a Finnish physics teacher discussing the nature of elegance with teenagers (*The Economist*, 2013a). With the help of these kinds of framings the Nordic context becomes associated with an effortless integration of, for example, environmental friendliness, climate neutrality, gender equity and inclusion, and knowledge-based economic development. Notably, these positive qualities are identical to the seven key principles articulated for social sustainability by Shirazi and Keivani as: 'equity; democracy, participation, and civic society; social inclusion and mix; social networking and interaction; livelihood and sense of place; safety and security; human well-being; and quality of life' (Shirazi and Keivani, 2017: 1537).

We suggest that the Nordic model achieves a degree of embedding of sustainable development goals via long-term association of values that matter in everyday life, across the social and political spectrum. That is to say that, in spite of political diversity within Nordic countries, traditionally there has been little need for argument that social, environmental, and economic sustainability goals and initiatives should be integrated. Social sustainability, in this way, meets the coherence test for residents of Nordic countries. More recently, as the Nordic model has risen in global prominence, external analysts such as international policy makers have also taken note of the value of this model, but here the value is of a different currency. Whereas, before the turn of the twenty-first century, the understanding and evaluation of urban policy were conducted primarily according to internal city goals and objectives, for global cities in our time, such internal evaluation is insufficient. The new currency for tracking the success and value of policy advances in sustainability is one of international recognition, which, in turn, establishes a city's policy actors as having standing to spread their model internationally. Blok describes this process in terms of 'policy mobilities' (Blok, 2012: 2334) and makes the case that Copenhagen specifically has achieved prominence for environment and climate policies on the international circuit. By achieving centrality in policy mobility terms, cities like Copenhagen set the bar for international practice in policies and urban design outcomes that are considered key to sustainability. Along with this perception, such cities are also ideally placed to export their eco-identity to policy actors such as planners, funders, and developers elsewhere, as they 'stage, in specific and different ways, a "cosmopolitan" sense of place-based eco-identity for "global" audiences and purposes' (Blok, 2012: 2328; see Rapoport, 2015: 114). In their presentations in city halls, media reports, conferences and trade shows, and international development reporting and ranking, place-based urban planning and development narratives get circulated in ways that often aim to increase the appeal of a singular model, internationally. It is worth asking whether older Nordic values, policies, and standards related to social sustainability have stood the test of international translation because they have achieved this prominence in environment and climate policy mobility circuits.

Those who would advance social sustainability today, and those who would advance a city on any platform, thus face a dilemma. Do they borrow the look, feel, and governance arrangements that are associated with successful work elsewhere, and risk that these may not be consistent with local values and habits, and so not work out in practice once contextualised locally?

Or, rather, do they craft a locally specific programme, and risk that the results of locally crafted advances will not be recognised as consistent with sustainability in the global mould and international understanding, and therefore potentially politically unsustainable? This dilemma comes into focus in cities in the Nordic region, where efforts to create social sustainability predate the use of the term 'social sustainability' in international policy dialogue, but the new international urban policy landscape imposes a different register of values and risks, costs and benefits of sustainability efforts against which cities are expected to perform. We will examine this dilemma in cases of Nordic eco-urban neighbourhoods in Sweden, Finland, and Denmark.

With certain origins in Nordic nations, an increasingly popular way of addressing problems of urban sustainability has been the creation of 'eco-urban neighbourhoods'. Eco-urban neighbourhood developments often involve innovative planning, design, and technological arrangements for living (Holden *et al.*, 2015). Eco-urban initiatives aim to address human needs in cities while simultaneously considering global climate change. Eco-urban projects are often portrayed as neighbourhood-scale urban models in which 'higher densities, brownfield sites, intensity of mixed use and social mix, more active discouragement of the use of private automobiles and more encouragement of active and public transportation' get showcased (Holden *et al.*, 2015: 11419).

Fashioning living examples of sustainable life opportunities in urban neighbourhoods can play an important role in improving social conditions. Although it makes sense that the primary beneficiaries of better social sustainability arrangements in eco-urban neighbourhoods are residents and occupants of these neighbourhoods, in terms of the race for global city status, the purported benefits of these eco-urban neighbourhoods do not stop at the local scale. Among other reasons for valuing these examples, Koch notes that such representations 'are important because interactions with the city are not limited to actually visiting' – images and stories are projected and repeated, and they serve to modify global urban imaginaries, or the ideas that people around the world have about that particular city, and about cities and urban life more generally (Koch, 2012: 2446–2447). Indeed, it would be difficult to justify the incremental public investment needed for model sustainable neighbourhoods if their beneficiaries were limited to the small number of people who live and work in them; the policy justification must include a broader base of benefits of these neighbourhoods as models for other places if they are to be considered in the public interest. An important question thus becomes whether it is possible for a model eco-urban neighbourhood to achieve both local and international qualities of sustainability, and social sustainability in particular. For, although these neighbourhoods may spread new images of positive development paths around the world, fostering a more globally consistent understanding of the urgency of sustainable development, the price for this result may be too high to pay if it comes at the cost of incorporating and facilitating local habits and values that differ from the current global best practice norm.

In this light, it is pertinent to ask what gets highlighted, and what gets downplayed, ignored, or even erased when creating eco-urban neighbourhoods and identities, and for what reasons. The answers to these questions have a major bearing on how model sustainable neighbourhoods will come to be understood in local and international contexts. Harris and Moore note, in relation to sustainable urbanism, that 'examples of "best practice", "industry standards" and technical checklists, [offer] a "global" model, aesthetic or brand that can be easily marshalled and mobilised in entrepreneurial attempts at fostering, legitimising and promoting contemporary forms of urban development' (Harris and Moore, 2015: 107). Faulconbridge considers how policy knowledge that is taken for granted by international professionals may 'devalue and write-out' local approaches in favour of 'generic global' approaches (Faulconbridge, 2015: 119).

Other researchers examining eco-urbanism have questioned the impact of international sustainability practices when they are applied in new places. China serves as a particular case in point, because top-down mandates mean that at least 230 eco-cities are currently in construction in that country (see Chang, 2017: 1724). Anna Hult has looked at how globally respected Swedish sustainability practices have been exported to Chinese cities in the form of eco-city models. Among the economic and political problems faced locally as these models are implemented in China, Hult notes that the eco-districts are designed and built to serve upper- to middle-class demographics and consumption habits, in spite of the fact that lower-class migrant workers live on-site already. In this light, she highlights how in fact these 'ecocities stimulate rather than mitigate inequitable development or mass consumption' (Hult, 2015: 550). This analysis spells potential disaster for the retention of the notion of equity in social sustainability within eco-cities being implemented in China.

Chang notes, in her research on the Chinese examples of Dongtan and Tianjin eco-cities, how actors have drawn not on existing habits and tradition but instead on international 'best practices' in their implementation (Chang, 2017: 1723). These model sustainable districts seek to form part of a 'globalizing city-making process' (Chang, 2017: 1720). Yang and Hu (2016) respond to concerns about this globalising city-making process in China by calling for a return to Chinese vernacular urbanism, which they call the Shan-Shui City, for more authentic sustainable urban development results on the ground.

Considering the evolving nature of sustainable urban planning and its promotion though international expertise (Rapoport, 2015: 114), the question we raise is: What narratives and models are driving the development of eco-urban neighbourhoods in the Nordic context, and at what cost to long-embedded values consistent with social sustainability? At issue in this question is, first, the degree of historical and policy continuity between approaches to integrate social sustainability into everyday life that have proven so popularly successful in the Nordic model (Figure 8.1), and the contemporary approaches being taken to embed social, environmental, and economic

Figure 8.1 Demands for the 'Nordic model' in Vancouver, Canada, 2017.
Source: photo by Meg Holden.

sustainability into model-sustainable neighbourhoods within Nordic countries. If the evidence from these new neighbourhoods demonstrates less continuity with local history and context, and more affinity for chasing a global ideal of the sustainable city that has yet to prove realizable, then these model eco-urban neighbourhoods are running the risk of losing ground on social sustainability terms, in the pursuit of profitability and legibility in the global urban policy circuit.

The significance of this question is not limited to Nordic countries. Rather, increasing numbers of countries look to the Nordics for answers to the global conundrum about adaptable policy, planning and design solutions for sustainability shortcomings, and the task of building new eco-urban neighbourhoods is one area where Nordic countries are forerunners. If model-sustainable neighbourhoods in Nordic countries are supplanting pre-existing characteristics that were conducive to social sustainability with features that disregard or threaten the role of social sustainability in urban solutions, there is a risk that social sustainability may have lost relevance as a vital component of these solutions. There is a risk that, if we do not answer this question effectively, the present and future potential of model eco-urban neighbourhoods worldwide will be compromised by a sense that, even in the lands where they originated, these models ultimately disregard the roots of social sustainability that pre-existed them, in pursuit of a globally transferable sustainability look and feel.

The analysis we offer in this chapter, to help answer this question, builds on a global scan of model eco-urban neighbourhood projects being undertaken in cities worldwide (Holden *et al.*, 2015). Drawing from this database, and from case studies conducted in 2015 in two such neighbourhoods, the first part of our analysis presents key motivations and elements of these two Swedish neighbourhoods, Hammarby-Sjöstadt, in Stockholm, and Bo01, in Malmö. This descriptive presentation will help to contextualise some of the key elements that have given rise to this neighbourhood-scale approach to urban sustainability, as well as some of the key elements contained within these prototype cases that relate to their prospects to advance social sustainability values. This context is presented in order to facilitate a more in-depth consideration of two other cases of eco-urban neighbourhoods. These cases draw from primary research conducted in 2015–2017, secondary policy, research, and media sources, as well as personal experiences, because they are led by the co-author native to the city in question. First, the case of Eco-Viikki in Helsinki: Finland provides evidence of contrast across the political, social, and cultural values enacted through everyday life in Helsinki and those implied by the design, branding, and marketing of the eco-district, a contrast that has expanded from the late 1990s to the present day. Second, the case of Nordhavn in Copenhagen, Denmark is presented as an emerging, large-scale district, sustainability initiative that embeds a degree of contrast with the social, political, and cultural values relative to social sustainability in Copenhagen as a whole. These cases are presented and analysed through a conceptual lens intended to bring into focus the precursor cultural, political, and social aspects of context that pre-dated these developments, on the one hand, and the elements and tendencies that have been built into their design and designation as model-sustainable neighbourhoods, on the other. This distinction allows us to consider the size and significance of any social sustainability gaps that are created in the offering of model-sustainable neighbourhoods in these cities, along with possible efforts to mitigate these gaps. We draw conclusions about the potential trade-off relationship between designing eco-urban neighbourhoods that meet local social sustainability needs, and designing such neighbourhoods to meet global city development demands.

Nordic origins of model-sustainable neighbourhoods

To introduce the aspiration to integrate model-sustainable neighbourhoods into the deep-seated social sustainability values that exist in the Nordic region, we draw upon the presentation of two precedent-setting cases in Sweden, namely Stockholm's Hammarby-Sjöstad and Malmö's Bo01. This is followed by a more in-depth treatment of model sustainable neighbourhood efforts in Helsinki and Copenhagen, and their implications for social sustainability outcomes.

Located on the canal at the south of Stockholm, Hammarby-Sjöstad is the largest sustainable urban development project completed in Stockholm to date. Completed in 2017, the district is designed to provide 11,500 homes for around 25,000–30,000 residents, as well as 10,000 jobs in a mix of office, light industrial and retail sectors (City of Stockholm, n.d.). The district contains a high residential density of about 270 persons per hectare, with primarily 1- and 2-bedroom apartments, as well as lesser proportions of studios, larger apartments, apartments for students, and assisted care dwellings for elderly people, i.e. it offers 'livelihood and sense of place' features, with an intent to contribute to human well-being and quality of life, for social sustainability by design.

The district was, before its redevelopment, an industrial harbour, characterised by slum-like qualities of residential areas and contaminated land (City of Stockholm, n.d.). The transformation of the area began in 1996 (Pandis and Brandt, 2011), catalysed by an Olympic Games bid. The district's environmental aim was that 'it should impose as little demand as possible upon resources, and be an environmentally well-adapted city district, whilst being at the forefront of international strivings towards sustainable development in densely populated urban areas' (City of Stockholm, 1996: 4, quoted in Pandis and Brandt, 2011). Although the Olympic bid was unsuccessful, it was the trigger for an extensive environmental programme which became a central element in the redevelopment of the area, based on sustainable use of resources, reduction of energy consumption, and maximisation of recycling. For residential dwellings, a specific goal was for district homes to have half the environmental impact of a standard, newly built home in Stockholm (City of Stockholm, n.d.). Other ambitious goals were set in the domains of transportation, water and drainage, building materials, land use, contaminated soils, lake restoration, and emissions reductions.

The Bo01 ('Living' in 2001, in Swedish) district was the first redevelopment phase of the Western Harbour, or Västra Hamnen, of Malmö. The district had similar characteristics to those of Hammarby-Sjöstad before redevelopment: industrial harbour in decline, contaminated and derelict site, within walking distance of the city centre. Redevelopment began as a showcase of the future for the 2001 European Home Expo. The total population of the Western Harbour in 2010 was 4,326. The mixed-use district also contains institutions such as Malmö University and employs approximately 9,000 people in around 260 businesses. Automobile traffic and parking structures are kept to the periphery of the district, with primarily walking, cycling, and small canal corridors in the interior.

Model sustainable neighbourhoods such as Hammarby-Sjöstad and Bo01 were never intended to model only environmental features. They were also intended to model new arrangements for living, working, and playing for all occupants and visitors. Some aspects of social sustainability, such as livelihood and sense of place, safety and security, social interaction, and well-being, were advanced through urban design and place-making programmes. Design features included mid-rise, medium-density residential buildings, with courtyard and walkable boulevard designs, common areas such as pocket parks, community gardens, and playgrounds, special features to promote physical activity such as bicycle facilities, waterfront promenades, and universal design for those with physical disabilities, transit accessibility, and access to the harbour.

These designs also featured new community art and facilities intended to promote civic participation, another key element of social sustainability. Hammarby-Sjöstad's GlashusEtt Information Centre provides an example. This impressive-looking modern building is intended as an interpretive centre for documentation of the area's environmental features including recycling initiatives and green building, and its systems-based approach to urban environmental design. As such, it has come primarily to serve visitors and study tours, whereas, for residents, this facility does not lend itself to regular use. More recently, as the district approaches full build-out, the City of Stockholm has made efforts to open GlashusEtt for consultation sessions with businesses and others. It remains to be seen whether this facility can form a fully functioning part of a home neighbourhood, for residents.

Whereas Hammarby-Sjöstad was motivated by a felt need to demonstrate 'world-class' sustainable urbanism to the world (and more precisely to the International Olympic Committee), Bo01 was placed also to meet a specific local need. Social sustainability is a term with considerable policy relevance in Malmö; the city launched a precedent-setting Commission for a Socially Sustainable Malmö, with a particular focus on equity, which has since expanded into a nation-wide Swedish effort. Within this context of concern for social sustainability gaps in the city, the first phase of the Western Harbour district was sited not just for the European housing exhibition, but also to provide a meeting place for segregated districts in this city of disadvantage. With a similar urban design strategy as described above for Hammarby-Sjöstad, the neighbourhood was intended as an attractive new waterfront area that would welcome urbanites of all kinds and promote the values of inclusion, social mixing, and diversity. The local press criticised the design: units included balconies considered to be ostentatious and large. However, the new neighbourhood proved popular among young professionals and families, who appreciated the location and proximity to work as well as nature reserves outside the city, and the larger units. It earned a positive local reputation despite its 'world-class' international design approach, not just for the opportunity to 'mix', but for the opportunity to 'live'. Maria, social sustainability researcher at the Institute for Sustainable Urban Development, says: 'the intention was that the apartments there are for maximum two children or couples, but a lot of families found it attractive' (in an interview with 'Maria', an employee of ISU Institute for Sustainable Development, Malmö, Sweden, 3 July 2015). The interest of families in living in these neighbourhoods came as a surprise to urban designers and planners in Malmö, much as was the case in Stockholm. This required design retrofitting to accommodate schools and day-care centres, the need for which had not been anticipated. In this respect, we might conclude that social inclusion of a fuller spectrum demographic occurred in these districts by surprise, rather than by design. How far does this experience advance social sustainability?

A fair assessment of the social sustainability outcomes of model-sustainable neighbourhoods depends on consideration of this interplay of global and local, visiting and living, and questions of demographic preferences. Can a neighbourhood be simultaneously open and accessible to the world as an example and a place that 5,000–10,000 residents uniquely call home? Can a neighbourhood intended as a model of sustainable design and planning also be responsive to local democratic demands for representation and decision-making? Goals and intentions related to such social sustainability principles as equity, diversity, democracy, and locality can be hard to discern within model-sustainable neighbourhood frameworks; indeed, they can always be expected to work out differently than planned on the ground and to work out differently over time and development stages, as well.

Behind every model-sustainable neighbourhood is a conception of public interest, whether implicitly embedded or explicitly expressed. Just who the interested public are, the interest they are expected to take in one another and in the spaces they occupy, and their relationships to the city and world surrounding their neighbourhood are built into the neighbourhood in important ways through the design vision, planning, policy, financing, and delivery of the new districts. Two cases of model-sustainable neighbourhoods in the Nordic cities of Helsinki and Copenhagen will serve as further elaboration of the contradictions confronted in Nordic cities as eco-urban development accelerates, and sometimes comes into conflict with longer-standing social sustainability attitudes, values, and practices. In Helsinki, we focus on the contradiction presented by a new district that provides much in the way of energy efficiency technologies, yet may undermine deep-seated cultural values of frugality and non-wastefulness in recent developments and marketing. In Copenhagen, we focus on the contradiction presented by a demand for social inclusion in the context of a model-sustainable district that struggles to justify the economic cost of social mixing. Eco-Viikki, in Helsinki, has been developed since the 1990s, and our other case, Nordhavn in Copenhagen, is newly inhabited and will reach build-out in 2050.

The case of Eco-Viikki, Finland

Eco-Viikki was the first ecologically designed neighbourhood in Finland. This 23-hectare area, situated in the southern part of Latokartano, in Viikki, is located about 8 kilometres away from downtown Helsinki (Aaltonen *et al.*, 1998: 7; City of Helsinki, 2017). The construction of this area lasted from 1999 to 2004, and the area is primarily residential (City of Helsinki, 2017).

The district of Viikki has a history of agriculture dating back to the fifteenth century (Luoto, 2004: 6), making it unique among surrounding Helsinki neighbourhoods. According to Luoto, cultivation here has been ongoing since the Middle Ages, and the old fields and buildings are a valuable part of the area's cultural history and identity (Luoto, 2004: 42). The University of Helsinki has had a presence since the beginning of the twentieth century, and it has used the properties and fields for agricultural and forestry education and research (see Luoto, 2004: 24), as well as cultural and historical celebration.

For example, every spring a public event is held in Viikki to release the cows out to pasture for the summer (Pajuriutta, 2017).

Eco-Viikki is also culturally situated within a wider distinctive context. Times of scarcity have left a deep imprint in the Finnish national psyche and, as Riikka Aro points out, the economic situation faced in post-war Finland helped maintain an already-entrenched and deep-rooted cultural tendency towards frugality and modest living (Aro, 2017: 18). In Finland, the two wars (1939–1940 and 1941–1944) left the nation relatively poor, and significantly influenced the country's development for decades. As Aro (2017: 18) notes:

> in postwar Finland [the] majority of people lived simply, frugally; there was not much to 'consume' and a great deal of materials were used economically, recycled, and reused. Thus, frugality, though not viewed in the same sense in contemporary discussions on sustainability, was the norm. The important notion here is that when others were viewed as living according to [the] same norms and ideals, it was more acceptable and thus easier for everyone to live in moderation. Frugality was also seen to serve the interests of the society and Finland as a 'nation'.

In addition to the post-war context, Finland experienced a deep recession throughout the 1990s. This came in the aftermath of the 1987 Brundtland Commission, and influenced the growing general awareness of ecological problems in Finland, colouring the national understanding of what sustainable development might look like, locally. Hakaste *et al.* (2005: 5) note, in relation to environmental awareness at the time, that 'There was a strong belief that by using technology and by developing specific methods one could achieve more "environmentally friendly" buildings and living environments'. At the same time, the recession created considerable pressure for eco-urban developments to be economically viable, because new development ideas for eco-neighbourhoods were drawn from Nordic countries among others.

It was within this wider context that Viikki was chosen to be a pilot area for an eco-community project in 1994, and the National Technology Agency of Finland (Tekes) significantly supported its development from the start. In 1998, the Finnish government started an ecologically sustainable building programme that provided a framework for many projects in Viikki (Hakaste *et al.*, 2005: 7); and the European Commission also provided financial support (Winkel, 2010).

As most buildings in Eco-Viikki were built at the same time in the early 2000s, there is no great variety in the age, design, or character of contemporary buildings. Eco-Viikki is primarily residential. It is surrounded by fields, the university campus, and a library. The residential buildings are symmetrically distributed, surrounded by carefully measured street lines, bus stops, and green plantings (Figure 8.2). The area looks and feels newly built and orderly.

Smart energy solutions have been highlighted as a particularly innovative part of Eco-Viikki (e.g. Winkel, 2010; Solpros, 2003). Furthermore, Viikki has

Figure 8.2 Eco-Viikki in the foreground.
Source: Google Earth, 2018.

been part of EU projects concentrating on energy consumption such as the PV-Nord project (Solpros, 2003). Invested developers, among others, frame Eco-Viikki as having 'pioneered sustainable construction techniques' in promotional materials (Skanska, 2008), using descriptions such as 'world-class sustainable solutions' and 'innovative solutions' (Tekes, 2011). *Helsingin Sanomat* (Helsinki's main newspaper) reported in 2001 that 'Eco-Viikki is the laboratory of ecological building' (in Finnish: 'Eko-Viikki on ekologisen rakentamisen laboratorio') (Salmela, 2001). Similarly, the City of Helsinki describes the area as 'a leading area of ecological building in Finland' (City of Helsinki, 2017).

Such notations and accolades, though ostensibly limited to the ecological aspects of model-sustainable neighbourhood development, are advanced for their purported contribution to social sustainability goals as well. The 2012 Nordic Innovation report, for example, refers to Eco-Viikki as making a pioneering contribution to 'enhance growth and welfare in the Nordic countries. The assessment is that the Nordic countries can and should be a pioneer in dynamic green and sustainable solutions' (Nordic Innovation Report, 2012: 8). This quote encapsulates the elision between the social sustainability goal of improving human well-being, and the global city development goal of creating solutions to sustainability challenges that are sufficiently 'world class' (Tekes, 2011) to be recognised and exported elsewhere, towards 'a "cosmopolitan" sense of place-based eco-identity for "global" audiences and purposes' (Blok, 2012: 2328).

'Pioneering' eco-urban construction is narrated as smart, innovative, and energy efficient in Eco-Viikki. The district's planning and development materials promote cosmopolitan urban values above the 'vernacular attachments to particular urban localities' (Blok, 2012: 2336). Older ways of living in the area, which encompass modest ways of living (see Aro, 2017: 18), Finnish values of frugality, and legacies of war and recession, are hard to find within the narratives and plans for this eco-urban neighbourhood. The elision of this rich historical and cultural context abandons the responsibility for the achievement of social sustainability to social memory. This is all the more obviously a need where the social memories themselves, of fields, farms, and more traditional agricultural ways of living, have obvious value for key social sustainability objectives from place attachment to civic participation. The story that begs to be told about Eco-Viikki is, instead, related to the 'world-class sustainable solutions' (Tekes, 2011) proffered on site, which are more amenable to purposes of international export and capital growth.

These dominant conceptions of Eco-Viikki, in sum, serve to distance this eco-urban neighbourhood from its local context and history. In so doing, they do a disservice to social sustainability. This further indicates a lack of understanding of local contexts and their significance to the planning and building of eco-urban areas. The implications here, in terms of social sustainability, are that important deeply rooted elements, which have formed a foundation for social values such as frugality, biophilia, and inclusion, do not fit the mould of world-class model-sustainable neighbourhood that Finland wants to be known for.

The case of Nordhavn in Copenhagen, Denmark

Initiated in 2005 by the Danish government and the Municipality of Copenhagen, Nordhavn (Figure 8.3) is one of the most ambitious large-scale urban development projects in Copenhagen (By & Havn, 2009). The land is owned by CPH City and Port Development (By & Havn Co., see www.byoghavn.dk/english/about/about2-uk.aspx), a public partnership between the municipality and the state.[1] The area represents the second phase of redevelopment of Copenhagen from the 1990s recession, following the Ørestad development. The strategic role of Nordhavn is to improve the attractiveness of Copenhagen for new residents and workers – 40,000 of each over a period of 40–50 years are the actual growth targets (By & Havn, 2009; Bisgaard, 2010; City of Copenhagen, 2015). The concept for the development, *Nordholmene – Urban Delta*, was the result of a 2008 international architecture competition, further elaborated into a sustainable development strategy, spatially arranged around small piers separated by water.[2]

As envisioned, Nordhavn should strengthen the city of Copenhagen's identity as a green capital (Nordhavn, 2017). To meet this goal, strategies have been created to promote the provision of renewable energy, more effective use of resources, and environmentally friendly modes of transport (see Mygind *et al.*, 2015). The strategy for Nordhavn explicitly combines ecological and social sustainability goals and entails the creation of both value and social diversity in a way that brings about a specific interpretation of social sustainability. This strategy states:

Figure 8.3 Nordhavn, 2018.
Source: photo by Majken Larsen.

> Nordhavnen will be a district of small islets close to the waterfront and a diverse, mixed city with room for everyone. As an urban development project, Nordhavnen spearheads efforts to improve climate conditions and show how cities can help reverse climate change without losing out on quality of life, welfare and democracy.
>
> (Nordhavn Urban Strategy, 2009: 2)

Nordhavn is located on a historical harbour that has been formed by landfilling over the last 150 years. Since the mid-nineteenth century, Nordhavn has experienced multiple land use changes due to industrial and technological factors in the shipping industry. Industrial activities have declined since the 1970s, and today only around half of the area is used for harbour-related activities (By & Havn, 2012). What brings a highly place-specific context to the district today is a small vernacular industrial harbour of fishermen, boat builders, and sailor associations in the area who have inhabited the area continuously since 1919.

The municipality of Copenhagen – where the Social Democratic political party traditionally has held power – has long worked with policies for promoting social cohesion and liveability (Bisgaard, 2010). This is reflected in a relatively large non-profit housing sector intended to provide affordable housing to a significant share of the urban population: the stated vision is for 20% of all housing to be non-profit. In recent years, this non-profit housing has been used as a major component of strategies to counteract urban segregation and accommodate diversity and social inclusion (City of Copenhagen, 2011, 2015).[3] Such longstanding ideals and policies for promoting social sustainability were also reflected in the vision for Nordhavn. However, critics have pointed out that these policies have sometimes served processes of gentrification and displacement of low-income groups (HT Andersen, 2017).

The basic rationale for development in Nordhavn is that the sale of building rights will finance the further expansion of the Copenhagen Metro (and thus ostensibly serve social sustainability goals in terms of providing greater access to urban mobility options). Thus the mandate of By & Havn is to operate outside direct public control, to make urban development decisions based on the principles of return on investment. Gaardmand (1996) has referred to it as 'the mahogany table method' of urban development due to its top-down governance structure. J Andersen (2003) has named it a politics of gambling with the public interest based on fluctuations in the market.

Nordhavn is envisioned as a vibrant city marked by diversity in urban life with a wide offering of activities and urban functions: 'The urban spaces, nature and not least the water shall provide for experiences, excitement and opportunities for expression for everybody' (Nordhavn, 2017). To promote diversity in use and different everyday life rhythms, the development strategy opts to use the ground floor of the buildings for retail whereas the upper floors are used for dwellings and offices (By & Havn, 2012). Furthermore, the vision accentuates goals of inclusion and equity by supporting a mix of different sizes, price ranges and forms of ownership of dwellings. The vision also notes that

> Nordhavn shall be a dynamic neighbourhood with environments that inspire new initiatives. There should be a broad range of institutions, shopping venues and experiences that form the basis for new jobs and initiatives. Nordhavn will thus help to strengthen Copenhagen as a significant, international knowledge-centre.
>
> (Nordhavn, 2017)

This could fairly be interpreted as an intention to support social diversity and avoid any trap of exclusivity.

At the same time, Nordhavn's visions of environmental and social sustainability are affected by, and dependent on, the economic rationality that frames the project. Concerning social mixing goals, home prices in Nordhavn are higher than for Copenhagen as a whole (Mygind *et al.*, 2015). Concerns raised about this conflict of values resulted in the 2014 National Planning Act, which allowed city governments to claim up to 25% of new housing in an area as non-profit. The first phase of Nordhavn to be developed, Århusgade, was built before the new Act, but the Municipality of Copenhagen has in fact worked to ensure the provision of some non-profit housing in Århusgade (Mygind *et al.*, 2015). Tight national control on local government budgets does, however, narrow the municipality's room for strategic long-term investments in non-profit housing, and thus its ability to affect the social mix of urban neighbourhoods. Mygind *et al.* (2015: 77) reflect that Copenhagen municipal government itself is the only developer able to act on the vision of social mix: 'it has been difficult to get investors into the idea of mixed building plots, and the Municipality of Copenhagen is still the only one who has tried to accomplish By & Havn's vision'.

Another initiative to counter the gentrification pressures in Nordhavn is to apply a 'creative zone' as defined in the 2005 Municipal Plan to a portion of the site. Creative zones are designations intended for former industrial areas, to support mixed use of older building stock by craftspeople and artists, with a maximum residential unit entitlement of one per enterprise. The notion is to preserve the existing character of older neighbourhoods even while adding new energy and enterprise, and to remove the pressure for profit from these specific zones. However, although such a zone has been designated in Nordhavn, creative zones elsewhere in Copenhagen have a low allowed density cap of 60% of the floor:space ratio, which does not apply in Nordhavn. This raises questions about how effective the creative zone may prove to be in alleviating the pressure to grow and profit.

Despite these efforts on the part of the Municipality of Copenhagen to address concerns about social sustainability limits in Nordhavn, the development process in this district stands out for a rather closed governance setup which leaves only limited channels of influence for the public. Blok and Meilvang (2015) also question the absence of effective public engagement in plan making in Nordhavn. They remark upon alternative voices being ignored, and the strategic use of sustainability to justify existing plans and actions:

> [S]ustainability has increasingly emerged as the cultural–political territory on which to achieve … settlements [and] the language of sustainability has come to be invoked as a new standard of evaluation, by which urban planners justify their future-oriented interventions and developments for the city.
>
> (Blok and Meilvang, 2015, p. 24)

Steeves (2017) notes how public engagement was part of the construction brief. But she also argues that the public must be engaged on a more regular basis and have the option to revisit the plans. This is even more important considering the long timespan of development for Nordhavn (Steeves, 2017).

Discussion and conclusions

The world began to take heed of the value of the Nordic Model in the 1980s, and this interest has grown over the past decades; learning from the Nordic model has been a topic for discussion at a recent World Economic Forum and in the pages of *The Economist*. *The Economist* (2013a) characterises the Nordic Model as 'a generous welfare state that does not cost the earth', particularly because this model has appeared to withstand the pressures presented in the 1980s by the neoconservative turn, and in the 2000s by growing global austerity, as well as the revanchist populism fuelled by fears of terrorism. Desirable features of the Nordic model include a high standard of living, a strong middle class, high employment rates, especially for women, and relatively low inequality and poverty. It appears, by many internal and external measures, driven by a wide range of philosophies, that the Nordic model is strong, successful, and consistent with the aims of social sustainability.

In considering the reputations of Nordic countries, accounting for local histories and analysing these cases, we voice concerns that practices of social sustainability being generated within these neighbourhoods do not strongly follow ideals that made the Nordic model successful in the first place. From the early model-sustainable neighbourhoods of Hammarby-Sjöstad and Bo01 in Sweden, Eco-Viikki in Helsinki, and most recently Nordhavn in Copenhagen, these were never intended to model only environmental features. They were also intended to demonstrate new arrangements for living and working in an inclusive manner that supports social as well as environmental and economic sustainability. Our cases from Sweden, Denmark, and Finland all provide evidence that social sustainability exists in different ways in different locations, and that this does not necessarily put the term at a conceptual disadvantage (Shirazi and Keivani, 2017: 1539). The same model does not work everywhere. If eco-urban neighbourhood developments aim to create prototypes suitable for building 'anywhere' they lose the ability to carry connections to local contexts, including local political settings, values, cultures, and identities. Planning eco-neighbourhoods in a similar manner worldwide narrows the possibilities to incorporate meaningful social sustainability ideals. When planning for future policies for eco-urban neighbourhoods, it is our contribution to highlight that social sustainability should indeed, read, look, and feel different in different locations. It is a further indication of the beauty of social sustainability that this idea takes form in place via multiple local variations and approaches across the globe.

With the help of urban design and community-building or place-making activities, the cases demonstrate an aspiration for the 'world class', sometimes in opposition to an understanding of local capacities, habits, and needs. With regard to social sustainability themes of inclusion, equity, access, or quality of life, the model neighbourhoods raise concerns about what happens when market objectives do not support social sustainability ideals. In a context of an underlying and longstanding culture that is consistent with social sustainability ideals, the risks that new neighbourhood models will instead generate exclusions and departures from longstanding social values must be considered. As evidenced by our cases, framings that aim to appeal to more globally recognised 'sustainability' understandings get circulated and repeated by invested funders, developers, and stakeholders, as well as by the city. Through media repetition, they become more normalised in everyday discussion and language (see Rapoport, 2015: 114). Eco-neighbourhoods themselves are formed through such framings, which further shape their social understanding. In this way, broadly disseminated urban planning and policy narratives can have powerful, sometimes amnesiac, effects on long-held social values in place which can further obfuscate longstanding Nordic social sustainability approaches.

In the case of Eco-Viikki, the sustainability ideals highlight the area's energy efficiency within a competitive international setting, using universal contemporary understanding of sustainability. This is understood and implemented in ways that support new innovations. At the same time, traditional ways of living with frugality and self-sufficiency, which have deep and meaningful roots in Finnish society, are not considered part of these new sustainability ideals.

In Denmark, the Nordhavn strategy specifically stated social sustainability aims with regard to social inclusion – ones that echo longstanding principles guiding urban development in Copenhagen. Nevertheless, the development shows modest results in implementing diversity and affordability for diverse publics, with national and municipal governments making back-stop efforts to retrofit policy frameworks in order to provide opportunities for local historical continuity and social mixing, without compromising the profit model of the district's development. Furthermore, alternative voices have been ignored by planners and developers, so that sustainability is used as a strategic rhetoric to justify specific development aims to enhance Copenhagen's green reputation.

Both Eco-Viikki and Nordhavn demonstrate the power of the shifting influence of economic and political pressures on social sustainability. Eco-Viikki was especially influenced by the recession in Finland and, in the case of Nordhavn, it has proven difficult to provide non-profit housing in areas with high real estate prices. If models of sustainability are being used to generate funds to finance other urban development goals, the risk exists that sustainability politics and planning supplant the political and social Nordic model with more neoliberal agendas, as can be seen elsewhere in the world. While this development is taking place, new formulations of eco-urban neighbourhoods and districts don't just make the local history invisible, but, in some cases, they obliterate that local and lived variety of social sustainability, by devaluing the local history on which it is grounded in favour of a different, global, capital-minded, sustainability ideal.

Considering Nordic successes, model-sustainable neighbourhoods demonstrate the true test of resilience of the 'Nordic model' in today's context of increasing diversity, inequality, and challenges to liveability and democracy. Our cases demonstrate that, in the process of creating contemporary eco-urban neighbourhoods, more vernacular attachments linked to specific local contexts often get ignored or even lost in the contemporary sustainability narratives. The creation of model-sustainable neighbourhoods in cases observed in Sweden, Finland, and Denmark demonstrates the risk that some ground for social sustainability is being lost. Brand-new conceptions around innovative energy solutions in Helsinki, and the creation of more exclusive housing options in Copenhagen, also harken to a faith in technological advances to solve persistent environmental and sustainability challenges. The local and unique contexts, cultures, traditions, and activities are hard to trace within the new sustainability initiatives in these cases. Conversely, traditional and older ways of living and social rule-making could potentially articulate with and enhance newer and broader understandings of eco-urban living and social sustainability.

Acknowledgements

The authors would like to acknowledge the research conducted by students at Roskilde University (Clara Mygind, Jonas Nikolaj Mertz, Katrine Ebdrup Damgaard, Nanna Lundgren Jensen, Sofie Wedum Withagen) and Simon Fraser University (Molly Steeves) that contributed vital insight to the research reported here. Financial support for this research was provided by SSHRC Insight Grant #435-2014-0465.

Notes

1 The Municipality of Copenhagen holds 95% of the share whereas the Ministry of Transport, Building and Housing holds 5%. The former mayor of the municipality of Copenhagen is managing director of By & Havn.

2 The municipality of Copenhagen, By & Havn, and the group of advisors (COBE, SLETH MODERNISM, Polyform, and Rambøll) took part in this process.

3 'The 2011 Municipal Plan makes it possible to develop areas of the city that lack general housing' (City of Copenhagen, 2011: Appendix no., p. 443).

> 'The greatest focus will be on areas with little general housing. [...] It will be e.g. Nordhavn, Carlsberg and Enghave Brygge [...] We have to secure a diverse and coherent city without ghettos whether it is ghettos of wealthy or socially marginalised people', says Morten Kabell, Mayor of Technics and Environment in Copenhagen.
>
> (Source 8: *Politiken*, 18 April 2015, our translation; Mygind *et al.*, 2015: 33)

References

Aaltonen T, Gabrielsson J, Inkinen R, Majurinen J, Pennanen A, and Wartiainen K (1998) *Ecological Building Criteria for Viikki*. Helsinki City Planning Department Publications: 6. Helsinki: City of Helsinki.

Andersen HT (2017) Copenhagen: The social costs of urban renewal. In: Cucca R and Ranci C (eds), *Unequal Cities: The Challenge of Postindustrial Transition in Times of Austerity*. Abingdon: Routledge, pp. 138–154.

Andersen J (2003) Gambling politics or successful entrepreneurialism? The Orestad Project in Copenhagen. In: Moulaert F, Rodriquez A, and Swyngdouw E (eds), *Globalized City. Economic Restructuring and Social Polarization in European Cities*. Oxford: Oxford University Press, pp. 91–106.

Andersen TM and Roine J (eds) (2015) *Whither the Nordic welfare model?* Special issue, Nordic Economic Policy Review 2. Denmark.

Aro R (2017) Living standards and changing expectations. Investigating domestic necessity and environmental sustainability in an affluent society. Jyväskylä Studies in Education, Psychology and Social Research 581. PhD thesis, University of Jyväskylä, Finland.

Bisgaard H (2010) *Københavns Genrejsning 1990–2010* [*The Regeneration of Copenhagen 1990–2010*]. Denmark: Bogværket.

Blok A (2012) Greening cosmopolitan urbanism? On the transnational mobility of low-carbon formats in Northern European and East Asian cities. *Environment and Planning A* 44(1): 2327–2343.

Blok A and Meilvang ML (2015) Picturing urban green attachments: Civic activists moving between familiar and public engagements in the city. *Sociology* 49(1): 19–37.

By & Havn (2009) *Nordhavnen urban strategy*. Copenhagen: By & Havn.

By & Havn (2012) *Nordhavnen. From idea to project. Inner Nordhavn.* Århusgadekvarteret. Copenhagen: By & Havn.

Chang I (2017) Failure matters: Reassembling eco-urbanism in a globalizing China. *Environment and Planning A* 49(8): 1719–1742.

City of Copenhagen (2011) Municipality plan. *Københavns Kommune,* Kommuneplan 2011.

City of Copenhagen (2015) Municipality plan. *Københavns Kommune,* Kommuneplan 2015.

City of Helsinki (2017) *Viikki.* Available at: http://en.uuttahelsinkia.fi/viikki/environment (accessed 12 November 2017).

City of Stockholm (n.d.) Hammarby Sjostad audio guide. Produced by GlashusEtt: Hammarby Sjostad's environmental information centre, City of Stockholm. Available at: www.onspotstory.com/no/audioguide/en/1175/Hammarby-Sjostad (accessed 14 December 2017).

City of Stockholm (1996) Hammarby Sjöstad environmental program. Stockholm: City Planning Administration, City of Stockholm.

Faulconbridge J (2015) Mobilising sustainable building assessment models: Agents, strategies and local effects. *Area* 47: 116–123.

Gaardmand, A (1996) *Magt og medløb. Om mahognibords-metoden og den korporative planlægning* [*Power and Opportunism. On the Mahogany Table Method and Corporate Planning*]. Denmark: Arkitektens Forlag.

Hakaste H, Jalkanen R, Korpivaara, A, Rinne H, and Siiskonen M (2005) *Eco-Viikki. Aims, Implementations and Results.* City of Helsinki, Ministry of the Environment. Available at: http://tempus.volgatech.net/presentations/eco-viikki_en.pdf (accessed 16 April 2017).

Harris A and Moore S (2015) Convergence and divergence in conceptualising and planning the sustainable city: An introduction. *Area* 47(2): 106–109.

Helliwell J, Layard R, and Sachs J (eds) (2017) *World Happiness Report.* Available at: http://worldhappiness.report/ed/2017/ (accessed 6 April 2018).

Holden M (2012) Urban policy engagement with social sustainability in Metro Vancouver. *Urban Studies* 49(3): 527–542.

Holden M, Li C, and Molina A (2015) The emergence and spread of ecourban neighbourhoods around the world. *Sustainability* 7: 11418–11437.

Hult A (2015) The circulation of Swedish urban sustainability practices: To China and back. *Environment and Planning A* 47(3): 537–553.

Iqbal R and Todi P (2015) The Nordic model: Existence, emergence and sustainability. *Procedia Economics and Finance* 30: 336–351.

Koch N (2012) Urban 'utopias': The Disney stigma and discourses of 'false modernity'. *Environment and Planning A* 44: 2445–2462.

Luoto R (2004) *Viikin Latokartanon tilan historia* [*The History of Latokartano in Viikki*]. Helsingin kaupunkisuunnitteluviraston julkaisuja [Helsinki City Planning Department], p. 15. Available at: www.hel.fi/hel2/ksv/julkaisut/julk_2004-15.pdf (accessed 14 December 2017).

Milne R (2014) Concern rises as cracks appear in Nordic model. *Financial Times* 2 November.

Mygind C, Nikolaj J, Mertz, Damgaard KE, Jensen NL, and Withagen SW (2015) Nordhavn: The new sustainable district in Copenhagen. Student report, Roskilde University, Denmark.

Nordic Innovation Report (2012) *Survey of Green Legislation and Standards in the Construction Area in the Nordic Countries, Background*. Available at: http://nordicinnovation.org/Global/_Publications/Reports/2012/2012_10%20Survey%20of%20green%20legislation%20and%20standards%20in%20the%20construction%20area%20in%20the%20Nordic%20countries.pdf (accessed 15 December 2017).

Nordhavn (2017) *Vision*. Available at: www.nordhavnen.dk/fra+vision+til+bydel/processen+hidtil/nh-vision.*aspx* (accessed 14 December 2017).

Nordhavn Urban Strategy (2009) *Nordhavnen. Sustainable City – The Copenhagen Way. Urban Strategy*. Available at: www.nordhavnen.dk/~/media/_newnordhavnen/nordhavnen_strategy_271009.pdf?la=da-dk (accessed 5 April 2018).

Organisation for Economic Co-operation and Development (OECD) (2014) *Income Inequality*. Paris: OECD. Available at: https://data.oecd.org/inequality/income-inequality.htm (accessed 6 April 2018).

Organisation for Economic Co-operation and Development (OECD) (2016) *The Nordic Welfare Model and the Refugee Crisis: Buffer or Weakness?* 3 May 2016. Paris: OECD. Available at: http://oecdinsights.org/2016/05/03/the-nordic-welfare-model-and-the-refugee-crisis-buffer-or-weakness/ (accessed 16 November 2017).

Pajuriutta (2017) Kevätlaitumelle pääsy on lehmille villin juhlan paikka – video näyttää eläinten riemun [Getting out to the pasture in the Spring is a significant moment for the cows – Video indicates the joy of the animals]. *Helsingin Sanomat* 6 May 2017.

Pandis IS and Brandt N (2011) The development of a sustainable urban district in Hammarby Sjöstad, Stockholm, Sweden? *Environment, Development and Sustainability* 13(6): 1043–1064.

Partridge E (2005) 'Social sustainability': A useful theoretical framework? Paper presented at the Australian Political Science Association Annual Conference, Dunedin, September. Available at: www.isf.uts.edu.au/publications/partridge2005social sustainability.pdf (accessed 4 April 2018).

Rapoport E (2015) Globalising sustainable urbanism: The role of international masterplanners. *Area* 47: 110–115.

Salmela M (2001) Eko-Viikki on ekologisen rakentamisen laboratorio. *Helsingin Sanomat* 4 June 2001.

Shirazi R and Keivani R (2017) Critical reflections on the theory and practice of social sustainability in the built environment: A meta-analysis. *Local Environment* 22(12): 1526–1545.

Skanska (2008) *Eko-Viikki, Finland*. Available at: www.skanska-sustainability-case-studies.com/index.php/latest-case-studies/item/157-eko-viikki-finland (accessed 15 December 2017).

Social Progress Imperative (2017) *Social Progress Index*. Available at: www.socialprogressindex.com/ (accessed 6 April 2018).

Solpros (2003) *Ekoviikki sustainable city projects*. Available at: www.kolumbus.fi/solpros/ekoviikki.htm (accessed 18 April 2017).

Steeves M (2017) Nordhavn – A case study. Unpublished working paper, Simon Fraser University, British Columbia, Canada.

Tekes (2011) World-class sustainable solutions from Finland. Available at: www.tekes.fi/globalassets/julkaisut/sustainable_solutions.pdf (accessed 14 December 2017).

The Economist (2013a) Welfare more for less. A generous welfare state that does not cost the earth. *The Economist* 2 February.

The Economist (2013b) The Nordic countries. The next supermodel. Politicians from both right and left could learn from the Nordic countries. *The Economist* 2 February.

Transparency International (2017) *Corruption Perceptions Index*. Available at: www.transparency.org/news/feature/corruption_perceptions_index_2017 (accessed 6 April 2018).

Winkel E (2010) *Ekoviikki: Sustainable City Projects in Helsinki*. Available at: www.eukn.eu/e-library/project/bericht/eventDetail/ekoviikki-sustainable-city-projects-in-helsinki/ (accessed 15 December 2017).

Yang Y and Hu J (2016) Sustainable urban design with Chinese characteristics: Inspiration from the Shan-Shui City idea [online]. *Articulo Journal of Urban Research* 14.

9 Social sustainability and new neighbourhoods

Case studies from Spain and Germany

Iqbal Hamiduddin and Marco Adelfio

Introduction

This chapter explores the interpretation and implementation of social sustainability principles into the design and development of new neighbourhoods using the case studies of Rieselfeld in Freiburg, Germany, and Polvoranca in Madrid, Spain. Through the implementation of community infrastructure and urban design features to facilitate social interaction, as well as a mix of housing styles and tenures to create a population mix reflective of the wider community, these neighbourhoods appear to embrace the traditional notion of the geographical or 'place' community. However, the relevance of the neighbourhood as a meaningful contemporary social entity has become less clear in recent years, because of the restructuring of lifestyle patterns and social relations over space, as a result of advances in transport and communications technology. Furthermore, urban development patterns have increasingly come to reinforce distance intensity through the separation of residential development from other urban functions and land uses, in both car-oriented suburban development and transit-oriented development (TOD) models.

Three fundamental questions provide the underpinning structure to this chapter. First, does the concept of the residential neighbourhood remain a relevant planning scale for framing and addressing social sustainability issues? The second section reviews the contemporary meaning and relevance of the neighbourhood as a scale for planning, against the 'hollowing-out' process that has occurred through processes of land-use separation and distance intensification. The second question asks to what extent are social sustainability outcomes realised in new neighbourhood schemes? Here we move to the two case study sites, in the third section, to consider the extent to which social sustainability principles have been addressed, by way of both secondary and empirical primary data during site visits detailed in the second section. The final question posed is how do housing delivery processes, including community-led planning and housing production, influence social sustainability? Here, the contention is that social sustainability is not always influenced just by the spatial factors that govern patterns of travel, land use, and activities, but longer-term processes of community engagement and resident input into the design, implementation, and upkeep of neighbourhood spaces can be important where such opportunities are presented. However, as discussed later in the chapter, housing tenure, development models, and the provision and management of public space all present significant opportunities and constraints to engagement in neighbourhood place-making.

Social sustainability and the neighbourhood

Within the literature the neighbourhood retains its importance as a social construct (Jenks and Dempsey, 2007) and an elemental social building block, serving as the setting for home life, socialising, growing up, retirement, and, for an increasing minority, a working life. Against this background it may be argued that neighbourhoods – the context for home life – are important social places deserving attention. Yet the relative importance of these spaces is not necessarily reflected in the level of attention that they have received, and the qualities of the spaces that have subsequently been produced. Indeed, Barton (2000: 49) notes that neighbourhoods have had a 'chequered' history in planning, being regarded as 'places' in their own right such as in Clarence Perry's 'neighbourhood unit' in the early twentieth century, before a shift in emphasis towards ease of access and movement by car in the following decades which has impacted on the social performance of the residential environment (Appleyard *et al.*, 1981). The question of spatial scale relates to both the geographical *extent* of the neighbourhood and the application of social sustainability principles – and particularly the idea of social mix. This pervasive scalar vagueness has previously been noted by Sarkissian (1976: 243), who asserted that 'throughout decades of discussion on social mix, remarkably little attention has been paid to the vital question of scale'. Therefore, although there has been considerable recent debate over the viability and desirability of policies to encourage social mixing at the neighbourhood scale (e.g. Andersson *et al.*, 2007; Galster, 2007; Galster and Friedrichs, 2015), the question of scale remains extant; a 'meso-level condition of urban space involving the composition of the resident population' (Galster and Friedrichs, 2015: 176) provides a simple working definition in relation to neighbourhood social mix.

The neglect of the neighbourhood as a physical and social entity reflects the broader underdevelopment of social sustainability in research and policy (Bramley *et al.*, 2009; Hamiduddin, 2015a) and only very recently have concerted efforts been made to bridge the gap in research (e.g. Dempsey *et al.*, 2012). A remarkable example is Montgomery's (2016) study of suburban housing developments in New Zealand, which combines social sustainability and social capital theories, highlighting the role of a locally based, intermediary community organisation connecting citizens with local authority, and a self-propelled resilience emerging as a reaction to natural catastrophes (in this particular case earthquakes).

In attempting to identify social sustainability outcomes in relation to physical features of the built environment, Shirazi and Keivani (2017: 1541) note that 'there is no single evaluation framework applicable to all disciplines and scales'. However, Dempsey *et al.* (2011) use social equity and sustainability of community as the two key dimensions of a conceptual framework, the former concentrating on matters of accessibility and mobility to amenities, opportunities, and people, and the latter on community cohesion – and with it better known concepts such as social capital – and long-term community viability. Dempsey *et al.* (2012: 94) suggest a close relationship between social equity and the distribution of amenities and opportunities, including employment, as well as the means to access – through transport or other communication forms. In relation to sustainability of community, Bramley *et al.* (2006: 5) identify the following five measurable parameters: (1) social interaction and networks, (2) participation in community groups and networks, (3) community stability, (4) pride or sense of place, and (5) safety and security. Lastly, on the matter of residential social mix, Hamiduddin (2015a) argues that the overall demographic profile of a neighbourhood community should broadly reflect the demographic balance of the locality, by age and ethnicity. We have attempted to operationalise these diverse but broad principles of neighbourhood social sustainability by way of indicators presented in Table 9.1, which we apply in an empirical investigation of two case studies later in this chapter. Although the framework has its limitations, we feel that it forms a logical progression from the indicators used in earlier studies (Dempsey *et al.*, 2012).

Neighbourhood models

Clarence Perry's neighbourhood unit model has provided a durable template for urban community planning over almost a century, incorporating many of the essential ingredients needed to build and sustain neighbourhood community life. Education and community facilities set in generous green space

Table 9.1 Social sustainability indicators

Factor	*Indicator*
Social equity	Public transport provision Employment opportunities Retail offer Education, health, and community infrastructure Green and recreation spaces Neighbourhood accessibility measures
Community sustainability	Social interaction Community participation Community stability Pride/sense of place Safety and security Demographic profile, relative to wider locale Housing access and affordability

form the core of the neighbourhood unit model. Retail is located on peripheral arterial roads, at the intersection with adjoining neighbourhoods where traffic is greatest. These arterial roads also direct through-traffic away from residential streets, helping to preserve the social amenity of these areas – and in turn social interaction and relations. However, it is self-evident that Perry's model was developed during the rapid and sustained motorisation of the 1920s United States. Thus, although the model permits convenient travel by foot across the neighbourhood, longer distance travel is predicated towards the private car. Related to these assumptions about long distance travel, employment-generating commercial land uses are absent from the model, which further assumes that a strict separation of different land uses must be maintained, as enshrined in the North American land use-zoning approach. Viewed through the lens of social sustainability, the original neighbourhood unit model remains relevant in the present day – notably in defending residential communities against the corrosive social impacts of motorisation. However, unless implemented with high-quality public transport provision, access to employment and social opportunities is problematic – most obviously for those with limited or no car access.

In order to address the social and environmental threats of motorisation that were being recognised in the 1980s (Appleyard *et al.*, 1981), a suite of 'compact city' models began to be developed which aimed to reduce car use and improve access by public and physical modes of transport. Perhaps chief among these, the TOD model was developed in the early 1990s by Peter Calthorpe (1993), as a 'moderate to higher-density development, located within an easy walk of a major transit stop, generally with a mix of residential, employment and shopping opportunities designed for pedestrians without excluding the auto' (Parker *et al.*, 2002: 3). Although TOD shares many of the features of Perry's neighbourhood unit, an obvious point of difference is the focusing of new development towards public transportation, to facilitate an overall reduction in car use and to ensure that residents enjoy a basic level of equity of transport and access. However, a rebalancing of transport to provide more equitable access does not necessarily lead to a balance in overall community. Some evidence indicates that processes of residential self-selection occur because of the lifestyle appeal of particular physical qualities of a residential environment, including access to transport modes (Handy *et al.*, 2006). Residential self-selection may influence travel behaviour, in turn, where residents holding particular attitudes, values, and demographic traits are drawn to particular environments that are oriented towards particular travel modes, for example a high-quality urban realm for walking or particular public transport modes (Mokhtarian and Cao, 2008). Taking TOD forward, the transit-oriented mixed communities (TOMC) model seeks to capture the perceived accessibility benefits of TOD while addressing the issue of demographic skewing through careful consideration of housing tenure and income spectrum, in order to provide a residential scheme that delivers 'truly affordable housing; stabil[e] transit ridership; broadens access to opportunity;

relieves gentrification pressures' (Center for Transit-oriented Development [CTOD], 2009: 3).

Finally, in Germany a distinctive variation of compact city planning described as the 'city of short distances model' combines the features of the TOMC with a focus on the co-location of employment-generating land uses with residential development (Feldtkeller, 2017; Hamiduddin, 2017). A stand-out example is Tübingen's Südstadt scheme with its twin focal points on the two former military areas of Loretto and the Französisches Viertel (French Quarter). The 1993 Südstadt master plan established an overall target for the creation of a 1:1 ratio of new jobs created per new homes constructed, with this 1:1 ratio applied specifically at Loretto and a more modest 0.4 target applied at the more suburban French Quarter site. This model replicates the dense and highly integrated mixed land uses found in historic city cores, as represented in Tübingen's Altstadt.

Implementation of the short distances model is dependent on planning systems being able to accommodate seemingly incompatible mixed uses such as residential and light manufacturing within a single block or building. Unfortunately many systems are not that flexible. For this reason the two case study schemes selected for evaluation later in the chapter are examples of new neighbourhoods that attempt to address a range of social sustainability concerns, but could be considered to be readily implementable under a wide range of planning conditions.

Methodology

Using the framework set out earlier in Table 9.1, the two case study schemes of Polvoranca and Rieselfeld were evaluated through a mix of documentary and empirical data analysis. These sites were selected because they are of comparable size and maturity, but represent different approaches to the application of sustainable neighbourhood principles. Social equity aspects were broadly evaluated in an objective manner using data from planning documents, and from site visits, which allowed the existence and relative accessibility of relevant neighbourhood features to be evaluated, including: employment-generating land uses, public transport networks, retail, education and community facilities, and recreation and green spaces. This element of research was limited by the fact that a subjective evaluation, in relation to the relative quality of these facilities, was not undertaken.

Community sustainability aspects, by contrast, did include a greater level of subjective evaluation in relation to factors such as pride and sense of place, or feelings of security. This part of the research was appraised both through documentary data to give employment and crime data, and through empirical primary data including household questionnaire surveys and semi-structured interviews of residents and policy managers conducted by the two authors between June 2010 and July 2013. Details of the data collected are summarised in Table 9.2.

Table 9.2 Empirical data collection undertaken at the two case study sites

Case study	*Questionnaire data*	*Interview data*
Polvoranca*	301 survey responses over 21 developments of which 43 persons in the category of 'public led mixed-use developments of the last 20 years', including Polvoranca	21 semi-structured interviews in 16 developments of which 3 in the category of 'public led mixed-use development of the last 20 years', including Polvoranca. Interviewees categorised as social, institutional, and economic actors
Rieselfeld	95 household survey responses were returned from 500 distributed, giving a response rate of 19%	5 semi-structured interviews were conducted as follows with policy managers and community representatives of Rieselfeld

Note
* The survey and interviews conducted in Polvoranca are part of a larger study analysing single-family housing, mixed-housing, and mixed-use development in the Madrid region.

The case studies: Polvoranca and Rieselfeld

The aim of this section is to use the social sustainability framework presented in the first section to explore the two case study schemes. Rieselfeld and Polvoranca share a number of basic similarities of scale and physical structure and layout. Both have approximately 10,000 residents – the target population presented by Clarence Perry in his earlier neighbourhood unit model – and both have generous provision of public space, and are well served by public transport. However, although the provision of public housing ostensibly forms a significant point of difference between the two case studies, the matter of housing affordability is addressed at Rieselfeld through the provision of cooperative and community self-build homes, as well as through municipal housing. This is reflected in a number of demographic characteristics of each resident population in turn.

Polvoranca[1]

The tradition of connecting suburban development with local rail networks in Madrid dates back to the nineteenth century. At that time, the Madrilenian historic tramway routes were already linking the city centre with its outskirts, deserving a protagonist role in Arturo Soria's project of Ciudad Lineal (Valenzuela, 1973). Contemporaneously, an initial outline of the suburban rail network, which still exists today, was traced, connecting the Spanish capital city with its surrounding towns and a set of second-home developments. At the end of the twentieth century, the practice of developing residential settlements around train stations was rediscovered and supported by regional policies, encouraging the deconcentration and diversification of non-residential (universities, shopping areas, hospitals, technological parks) and residential uses in suburbia (Arpegio, 1993). With regard to housing, public administration has fostered the development of mixed-housing and mixed-income schemes, including a remarkable presence of subsidised housing, with the aim of promoting social integration and cohesion. In some cases, the central role of public transit in such public-led schemes led to the construction of suburban train stations, producing the Madrilenian autochthonous version of the TOD pattern. The selected case study, Polvoranca, in the municipality of Leganés, represents the evolution of a former working-class dormitory town situated south of Madrid into a mixed-use post-suburban centre, fostered by the aforementioned public policies for land-use diversification (Table 9.3). Such a process can be linked with a socioeconomic upgrade of local residents (Adelfio, 2014). The choice of a suburban development for a social sustainability study advocates for 'a more open debate on the integration of social sustainability in suburban environments', still impregnated by a stereotyped anti-urban 'negative connotation' (Adelfio, 2016, p. 4).

Polvoranca's master plan dates back to 1999. Arpegio, a public agency of the regional government (Comunidad de Madrid), was in charge of land management and obtained the land by means of compulsory purchase orders. In search of consensus, 'the regional government signed a series of agreements in order to give back more than 70% of land to the original land owners after expropriating and urbanising it' (De Blas, 2004: 114).

Social sustainability is shaped through the inclusion of a wide set of qualities and features (Table 9.4). The inclusion of subsidised flats has favoured the settlement for young families and first-time buyers (more than 50% of residents are aged between 25 and 49) along with terraced-home owner–occupiers. An interviewee from a residents' association declared that 'young families with children feel integrated into the neighbourhood'. Urban design was expected to encourage social interaction through a pedestrian-friendly environment (including two commercial streets, a mixed-use square, shared-surface residential streets, high internal street connectivity with an average block perimeter of 289 m. and public facilities at walkable distance). Approximately 15% of the master-planned area is devoted to green and local public spaces (Figure 9.1). Although it contains 11.57% of pedestrian streets

Table 9.3 Key features of Polvoranca

Parameter	
Development concept	Sustainable urban extension, built around a suburban train station
Total area	131.83 Ha
Number of dwellings	3,600 (1,200 houses; 2,400 subsidised flats)
Gross density	27.3 dwellings/Ha
Population density	78.4 pp/Ha (2017 population in municipal register)
Layout	Grid with internal shared-surface streets, two mixed-use commercial streets, and public main square

Table 9.4 Social sustainability outcomes at Polvoranca

Factor	*Indicator*	
Social equity	Public transport	Suburban train station, providing direct and reliable service to Leganés and Madrid. Bus service, less frequent than train Train ridership (percentage of residents in 2013): 30% of total residents; 50% of residents aged 15–64 Percentage employed residents travelling to work by private/public transport: 52%/32% in 2011
	Employment opportunities	Mostly small-to-medium size employers providing neighbourhood services: 72 economic activities in 2017 (according to the Madrid Regional Institute of Statistics) of which: 44 with 3–9 workers, 23 with 10–49 workers, 3 with 50–99 workers, and 2 with 100–249 workers
	Retail offer	Grocery stores, pharmacies, banks, bars, and restaurants
	Education, health, and community infrastructure	Day-care centre for elderly people, community centre, nurseries and primary schools, sports facilities
	Green and recreational spaces	The master-plan devotes 14.58% of total planned area to green and local public spaces; 19.23 m^2 of green and public spaces/inhabitant (considering a population of 10,000 inhabitants estimated in the master plan)
	Neighbourhood accessibility measures	'Barrier-free' public facilities set in the master plan Pedestrian and shared streets included in the master plan 27,676.3 m^2 of pedestrian streets (11.57% of Leganés) (Ayuntamiento de Leganes, 2010) 4,500 m (length) of bike routes (Leganés: 35,800 m) (Universidad de Alcalá, 2010) Maximum distance to basic public facilities (educational, social, sports) is 500 m, established in the master plan. Polvoranca is accessible by suburban train with a 5-minute frequency in morning peak hours; 20-minute journey to Madrid central station

Community sustainability	Social interaction	Use of eating places and outdoor spaces as inherent in Spanish culture (Herzog, 2010) From interviews: high level of interaction during weekends and late afternoon, especially in the main squares/boulevards and bars
	Community participation	From interviews: sense of community and community activities for families through residents' associations
	Community stability	Polvoranca is meant to introduce a socioeconomic upgrade in Leganés (Adelfio, 2014). Less than 5% are non-Spaniards, according to municipal statistics
	Pride/sense of place	Described as 'suburban area' in interviews and as 'mainly residential neighbourhood' in the Leganes plan for sustainable mobility (Ayuntamiento de Leganes, 2010: 286). Sense of attachment not specifically detected
	Safety and security	Sense of security perceived positively in interviews. No statistics available at the neighbourhood scale
	Demographic profile, relative to wider locale	Average household size 2.9 (Leganés: 2.6; Madrid municipality: 2.4) in 2011 Percentage aged 6–18 years old: 19% in 2014 Percentage aged 65 years+: 3.6% in 2014 Mean age of residents: 30 (Leganés: 41; Madrid municipality: 42.8) in 2011 Percentage people in employment: 50% in 2011 (excludes students, children, retired, unemployed); Leganés: 41% 96.84% inhabitants with Spanish nationality (2017 municipal statistics); Leganés: 89.78%
	Housing access and affordability	2,400 affordable homes/3,600 in total

Figure 9.1 Polvoranca's main square.

Source: photo by Marco Adelfio.

(Ayuntamiento de Leganes, 2010) of the entire municipality, the deficit of barriers to impede access of vehicles, together with an insufficient pedestrian flow, favours their misuse as 'thoroughfares' and 'increases their level of danger' (Ayuntamiento de Leganes, 2010: 286). The use of cobbled streets for pedestrians may hinder the passage of wheelchairs and children's buggies (Ayuntamiento de Leganes, 2010). The neighbourhood is also provided with ad-hoc bike routes, representing 12.57% of total Leganes bike routes (Universidad de Alcalá, 2010).

The use of public spaces is especially relevant in the Spanish context and inherent in the Spanish culture (Herzog, 2010). As 'critical devices', they often create a 'permanent outdoor urban salon' with 'tables and chairs parade for kilometres' (Herzog, 2010: 34). The objective of creating a vibrant and lively public area is only partially achieved because 'there are no offices. Life begins once you are back from work' when the 'families gather round the square' (interviewee: president of residents' association). The need to travel for local inhabitants has been partially tackled in relation to basic shopping and amenities, which are well used. Most of residents (approximately 85%[2]) already acquire basic goods within Polvoranca. For other purposes, they 'mostly use the train, then their cars and then local buses as their last option. You can reach the centre of Leganes in 3 minutes and the centre of Madrid in 25 minutes by train' (interviewee: member of residents' association). According to the sustainable mobility plan of the municipality of Leganés, the inhabitants 'commute by car or public transport to the shopping areas and amenities of Leganés' (Ayuntamiento de Leganes, 2010: 286). In spite of this, the master-plan document of Polvoranca states that

> the role of facilities is mutable, as they are satisfiers of needs. Hence, their use will change with time, adapted to the requirements of social changes. New needs and the rise of emerging groups in the city require [...] facilities for both past and future social aspirations, which are impossible to estimate at the moment.
>
> (Arpegio, 1999: 34)

According to 2011 Census statistics, only 7.5% of residents in employment are working at a walkable distance (less than 10 minutes). Despite most employed residents commuting to work by car/motorbike (52%), more than 30% use public transit. The use of private vehicles for work commuting is lower when compared with three other suburban mixed-use developments in former suburban dormitory towns around Madrid.[3] Moreover, according to the 2011 Census, considering that 50% of population is employed, 10% unemployed, and 40% inactive (Leganés values are respectively 41%, 14%, and 45%), we may suppose that public transit is also used for other reasons, as demonstrated by the number of train users, measured through the number of outbound journeys (Consorcio Regional de Transportes de Madrid [CRTM], 2013) from the local train station and displayed as a percentage of residents (50% of those aged between 15 and 64). For all the aforementioned features, Polvoranca may be considered to be a partially successful attempt to create a mixed-use, transit-focused development, accurately planned and designed for the needs of daily life, but probably not intended to be completely self-sufficient. The deficit of offices is a characteristic of most public-led, mixed-use developments created during the last 20 years in the Madrid region, and must be considered as a necessary improvement for the future, to provide these places with real 24-hour vitality.

Rieselfeld

Spurred by the energy crises of the 1970s and more recent concerns over global climate change, the small city of Freiburg has recently been dubbed Germany's 'green leader' (Buehler and Pucher, 2011), having encouraged a doubling in public transport use during the 1980s, expanded its cycle network by almost tenfold over three decades since the 1970s, and introduced car-restraining measures across the city. The city's population has doubled since the 1950s and continues to grow from a present figure of 215,000, owing largely to former students who remain after graduating from the city's prestigious university. A long-term housing shortage led planners to create two new suburban development schemes in the early 1990s, at Vauban and Rieselfeld, on the city's westerly fringe. Rieselfeld was designed for a resident population of 10,000 in 3,700 homes (Table 9.5), with a high degree of autonomy in mind through a full range of services and amenities.

Clarence Perry's community model is reflected in the layout of Rieselfeld, where peripheral distributor roads direct through-traffic away from residential

Table 9.5 Key features of Rieselfeld

Parameter	*Data*
Development concept	Sustainable urban extension, built around a tram extension
Total area	70 Ha
Number of dwellings	3,700
Gross density	53 dwellings/Ha
Population density	132 people per hectare
Layout	Grid with car-free superblocks

streets, and community facilities, including schools, library and inter-denominational church, create a municipally oriented core of the scheme. The development has been designed with traffic reduction in mind from the outset, with the grid-form layout of residential streets creating car-free blocks, and although the car-parking ratio is relatively high, at 1.5 spaces per home, a substantial proportion of residential parking has been allocated in subterranean car parks. The tramline has created the central Rieselfeld Allee axis through the development, and a significant retail offer, including three supermarkets and a wide range of smaller outlets including banks and a post office (Figure 9.2). Interconnecting bus services to outlying villages make Rieselfeld an interchange node, generating additional footfall for local retailers, including a twice-weekly market held in a public square adjacent to the church and community hub.

Rieselfeld's scale means that it can provide the spectrum of amenities and retail outlets to service the resident population, while the barrier-free design of the neighbourhood means that universal accessibility to these are assured (Table 9.6). A significant amount of new employment has been created, within the central retail and services hub, and also in the edge-of-development business park. For longer distance travel into the city, all homes are located within 400 m or a 5-minute walk to one of the three tram stops within the development, with each offering step-free access on to the tram vehicles. As a transport alternative, Rieselfeld is connected by dedicated cycle routes into the city, and each residential block and house features generous and secure cycle storage. Green spaces have been generously provided throughout the scheme, offering a mix of formal gardens or recreational spaces, as well as less formal 'wild' spaces that allow children to play in relative seclusion, but close to home. In terms of social equity – as defined in terms of accessibility – Rieselfeld performs well because a diverse range of needs is met within the locality, and also because residents can move easily within the neighbourhood and out of it without the need to use the private car.

Analysis of travel behaviour from the household questionnaire data (Hamiduddin, 2015a, 2015b) shows a high level of localisation for access to essential needs and services, for example with 32% of residents walking to local shops in the neighbourhood for regular grocery shopping and a further

Figure 9.2 Central Rieselfeld.

Source: photo by Robin Hickman.

34% cycling to shops within the general locality. This pattern is consistent with the high level of retail provision at Rieselfeld, which includes two supermarkets – itself reflecting Freiburg's policy of encouraging smaller, neighbourhood-scale grocery stores through an 800-square metre store size limit across the city. Planners have been rather less successful in their attempts to minimise work travel and, although in principle one job has been created per ten residents, the proportion of residents working within the neighbourhood is below 5%, indicating a spatial work–home imbalance. The development approach has succeeded in producing sustainable patterns of travel behaviour generally. Although the ratio of household car ownership (72%) mirrors the overall city average (74%), only a quarter of residents regularly drive to work compared with half of residents who cycle to work on a regular

Table 9.6 Social sustainability outcomes at Rieselfeld

Factor	*Indicator*	
Social equity	Public transport	All homes within 400 m of a tram station, giving direct and reliable service to centre and to key destinations throughout city. Bus services serve suburbs and outlying villages Car ownership: 72% (Freiburg 74%) Journey to work by car: 26% (Freiburg 17%)
	Employment opportunities	A business park containing light industry and offices is located on the edge of the scheme. Overall, 0.27 jobs have been created per household
	Retail offer	Grocery stores, restaurants, cafes, twice-weekly market
	Education, health, and community infrastructure	Full range of schools, library, police station, medical centre, and community organisation
	Green and recreation spaces	Generous provision of green space, including communal gardens within each housing block, extensive public gardens, and an adjoining nature reserve
	Neighbourhood accessibility measures	The neighbourhood is designed to be 'barrier free' with fully accessible public spaces and residential buildings

Community sustainability	Social interaction	Average number of persons known in neighbourhood = 64 (control neighbourhood = 22)
	Community participation	65% believe the scheme has a moderate or strong sense of community (control neighbourhood = 22%)
	Community stability	Low housing turnover – characteristic of the city
	Pride/sense of place	60% feel a moderate or strong sense of belonging to the neighbourhood (control = 20%)
	Safety and security	
	Demographic profile, relative to wider locale	Total population/9.384 (2012) Average household size/2.54 (Freiburg 1.84) Percentage aged 6–18 years old: 21 (Freiburg 10.2) Percentage aged 65+ years: 6.1 (Freiburg 16.3) Percentage receiving state support: 9 (Freiburg 11.4)
	Housing access and affordability	20% public housing stock 10% group self-build (*Baugruppen*)

basis. Only 17% regularly use public transport to access work, reflecting Freiburg's relatively small and 'cycle-able' size.

In terms of community sustainability, it must first be noted that Rieselfeld's population is heavily oriented towards younger families, compared with the overall Freiburg average. However, as one interviewee – a former director of planning – noted, this demographic skewing reflects a long-term housing shortage caused by an expanding younger population of university graduates and migrants that this scheme – and others elsewhere across the city – were designed to address. Although notably skewed, Rieselfeld's population is better balanced than other new residential schemes in the city, including the famous Vauban new quarter, where there are few older residents. Approximately 20% of Rieselfeld's housing can be labelled as 'permanently affordable', being provided by the state or through cooperative groups, while a further 10% of housing has been delivered through group self-build or '*Baugruppen*' schemes. The latter provide a significantly greater level of affordability for the first wave of resident builders because of initial cost savings over comparable developer-built housing (Hamiduddin and Gallent, 2016), but these are not maintained in perpetuity, and the value of a *Baugruppen* home will track the market. As a consequence, covenants have been applied to some *Baugruppen* schemes to restrict resale within a specified period – often five years – in a bid to prevent speculative developments being created by individuals or organisations in search of a quick profit. As *Baugruppen* homes tend to be more individually designed they both tend to enhance the character of their immediate locality and, because of the active input required of households into the design of their home, can offer an avenue for binding people to their home environment.

Discussion

Both case studies fall into the categories of new neighbourhoods and TOD at the same time. Although the TOD concept embraces environmental sustainability goals by encouraging public transport alternatives to car use, social sustainability principles have often been side-lined, reflecting an historic vagueness surrounding the term that has only been recently sharpened by a new structure provided by social justice principles under the heading 'social equity', as well as the ability for a community to regenerate itself, lent by 'sustainability of community'. Rieselfeld in particular has achieved a modest success both in creating a more representative demographic balance than other new TOD schemes in the city and, in particular, in marrying a relatively more socially 'balanced' resident community with environmentally sustainable patterns of travel.

Polvoranca's best achievements are related to social integration and the creation of a sense of community. The differentiation of prices within subsidised housing has partially balanced its concentration in multi-family buildings, improving social equity. The provision of local amenities and public areas, the organisation of activities by the residents' association, together with the Spanish culture of socialising in outdoor spaces, have favoured social interaction. Polvoranca would have benefited from a broader mix of uses and it is affected by a sense of physical isolation from the Leganes' urban continuum (arterial roads acting as barriers). At the present time, this sort of public-led project requiring an extensive use of compulsory purchase orders and significant public expenses would not be viable. Fortunately, Polvoranca has achieved a higher degree of completion than other contemporary developments, including essential amenities for daily life and a reliable public transit service that is not available in other Madrilenian, mixed-use, non-TOD schemes.

Polvoranca and Rieselfeld have been developed as urban extensions to address the rather different housing market needs of Madrid and Freiburg, respectively, the former to accommodate a high level of social housing targeted towards the unemployed in a country where the public housing sector has historically been rather small by international standards. On the other hand, Rieselfeld reflects Germany's overall housing market structure, with the bulk of homes for private rent, a smaller proportion for sale, and approximately a third as affordable homes – through either public housing or private *Baugruppen*. The age profile of residents is similar across both developments, with a large representation of younger residents and families reflective of local housing demand – particularly in Freiburg. Although different in the design and planning detail, both schemes have addressed many aspects of social equity and community sustainability, through:

- The provision of local services, amenities, and a range of outward transport connections;
- The creation of streets, public spaces, and communal environments conducive to social interaction;
- A mix of housing types and tenures, to encourage a mix of different social groups.

There are a number of other aspects to reflect upon, with implications for future practice. Foremost is the need to reconcile an intrinsic presumption of travel – for employment as well as broader social participation – which appears firmly entrenched within the TOD model, with the development and maintenance of social capital. Deconstructed, the two case studies may contain many of the fundamental ingredients necessary for supporting social sustainability, but both are also largely satellite suburbs where daily commuting is a requirement for most working residents. The mixed-use living quarter concept builds on the basic high density, highly permeable, car-reduced block structure that forms the development pattern at both Polvoranca and Rieselfeld, and applies a mix of primary residential and employment-generating land uses across the greater part of a scheme, in order to create both greater levels of self-containment in living and working, and daytime activity that helps to support secondary retail and services.

'City of short distances'

Rieselfeld is sometimes cited as an example of 'city of short distances' planning (Daseking, 2015), an urban compaction strategy rolled out in Germany since the late 1980s to reduce the 'distance between living, working, [...] so as to avoid motor vehicle traffic and at the same time increase the attractiveness of public transport, the bicycle and pedestrian usage' (Wegener, 1994: 9). The Spanish case study of Polvoranca, as a mixed-use neighbourhood, includes some of the expected benefits of such a model. However, because only modest levels of new employment have been created, most residents of both case studies still need to travel out of their neighbourhood to work.

Better examples in Germany include the edge-of-city Loretto and French Quarter areas of Tübingen's southern fringe or Südstadt scheme, where job:household ratios of 1:1 and 0.4 have been achieved through planning ordinances that require the street frontage of all buildings to be dedicated to non-residential uses, and permission for light industry and commercial land uses to be integrated in otherwise residential blocks (Hamiduddin, 2017). The effect of the policy has been a proliferation of small businesses and organisations to Loretto and the French Quarter, where just over a third of employees both live and work in the neighbourhood, compared with 22% in a control neighbourhood (ibid.).

In relation to social sustainability, local employment can be important for those who might otherwise be excluded from working – those caring for others, for example – and it can help to cement and sustain local social networks. However, almost as important are the indirect or secondary effects of the daytime footfall from employees within the neighbourhood, who, in turn, help to sustain a significantly greater diversity of retail and services than would otherwise be possible, serving to strengthen neighbourhood social life further.

Neighbourhood production

In Rieselfeld approximately 10% of residents could help produce their own homes through collaborative self-build or *Baugruppen* development. At Tübingen's Loretto and French Quarter areas, this mode of delivery has been used to provide the bulk of new homes and buildings – both in response to demand and also because mainstream developers viewed the production of mixed-use quarters as being too risky a venture. At Rieselfeld, as at the Tübingen schemes, the *Baugruppen* approach has created a diversity in the built environment that can be regarded as a 'new piece of the city' (Feldtkeller, 2015) rather just 'another' housing scheme. The process of housing co-production can also help catalyse strong social relationships between residents (Hamiduddin and Gallent, 2016), whereas the heightened sense of ownership and belonging to a locale can help with the upkeep of buildings as well as the maintenance of social relations. It follows that many collective self-build schemes present a canvas for the energy, place-making instincts, and shared interests of a community of self-builders. In contrast, the development of Polvoranca followed mainly a top-down approach, as part of regional strategies (Arpegio, 1993), using expropriation to acquire land ownership (Adelfio, 2014). Strongly dependent on public investment, a project of this size probably would not have been viable after the emergence of the global economic crisis.

Conclusions

This chapter began by reviewing the meanings of social sustainability in relation to the neighbourhood and by questioning whether the neighbourhood remained a meaningful scale of planning to consider the twin 'pillars' of social equity and sustainability of community. Having satisfactorily concluded that the neighbourhood retains currency both in defining a place community and as a spatial scale – as Clarence Perry depicted in the 1920s – a framework was then applied to two case studies that were developed as good practice examples. The overall conclusion from the case study investigation is that there is much to commend about Polvoranca and Rieselfeld, particularly with respect to attracting a range of resident groups, providing a basic 'safety net' of amenities and services within the neighbourhood, and good public transport access to their respective wider urban areas. Polvoranca arguably represents a stronger example of social mix and inclusion of the diversity of social groups, whereas Rieselfeld's structural orientation towards its mixed-use central axis and tramway system makes it arguably the better in terms of access to both local services and opportunities, and the wider city. However, we note that our conclusions are based on a limited set of indicators, informed by a theoretical framework that must undergo further development and testing.

In order to move the policy discussion on socially sustainable neighbourhoods forward, we believe that there are two important aspects for further practical consideration: first, the relative balance of primary uses within the neighbourhood. Like many new neighbourhood schemes, Polvoranca and Rieselfeld are predominantly residential areas providing only limited land-use mixing and limited employment opportunities. Elsewhere in Europe, the historic separation of land uses has begun to be challenged. Tübingen's Loretto and French Quarter schemes demonstrate that integrated land-use mixing is both achievable and potentially desirable from both an environmental and a social sustainability perspective (Hamiduddin, 2017). Given the long-term tendency towards greater home working and decentralisation of manufacturing, residential design should aim to better facilitate and support neighbourhood working. Our first policy recommendation is therefore that new neighbourhoods should include a diversity of land uses, with provision for a high level of local employment.

The second aspect is the delivery of new housing and the potential for encouraging diversity in supply, as applied at Rieselfeld through group-build or *Baugruppen* housing. As a developing area of research, social sustainability in relation to the built environment perhaps warrants greater consideration of production processes. There are many reasons to support a diversity of housing supply, with alternative models such as self-build, group-build, and co-housing encouraging greater affordability, the development of social relations, and physical diversity of the built environment leading to stronger identity as 'places' (Hamiduddin, 2015b). Our second policy recommendation is therefore that new neighbourhoods should be places of built diversity that reflect a broad social mix. We should strive to create diverse, inclusive, and enriching neighbourhoods that nourish social relations between residents, and reaffirm a sense of belonging.

Notes

1 The area is given different names in its master plan ('Residencial Polvoranca') and in municipal statistics ('Arroyo Culebro'). 'Polvoranca' is the name used for the purposes of this chapter.

2 Estimated value extracted from a survey related to 301 residents living in 21 selected developments of the Madrid region. The value of 85% is referred to three public-led, mixed-used developments of the last 20 years, including Polvoranca. Therefore, it must be intended as an estimation and not as a statistically significant value.

3 *Loranca Garden City* in Fuenlabrada (55%), *Soto del Henares Eco-neighbourhood* in Torrejón de Ardoz (56%), and *Pórtico del Sur* in Getafe (60%) are all provided with metro or train stations, although Soto del Henares was still a work in progress when this research work was developed.

References

Adelfio, M. (2014) Single-family housing and changing social profiles in former working-class areas: Madrid's south-western suburban ring as case study. *Journal of Urbanism* 7(3): 261–285.

Adelfio, M. (2016) Socially sustainable suburbia: Addressing a puzzling and multi-faceted issue. *Articulo – Journal of Urban Research* [online], 13. Available at: https://articulo.revues.org/2928 [accessed 19 November 2018].

Andersson, R., Musterd, S., Galster, G., and Kauppinen, T. M. (2007) What mix matters? Exploring the relationships between individuals' incomes and different measures of their neighbourhood context. *Housing Studies* 22(5): 637–660.

Appleyard, D., Gerson, M. S., and Lintell, M. (1981) *Livable Streets, Protected Neighborhoods*. Berkeley, CA: University of California Press.

Arpegio (1993) *Madrid Gran Sur metropolitano. Territorio, desarrollo regional y medio ambiente* [*Madrid Metropolitan Great South. Territory, Regional Development and Environment*] (Spanish/English). Madrid: Comunidad de Madrid.

Arpegio (1999) *Plan Parcial de Ordenación Sector PP5*. Madrid: Comunidad de Madrid.

Ayuntamiento de Leganés (2010) *Plan de Movilidad Urbana Sostenible* [*Sustainable Urban Mobility Plan*]. Leganés Town Hall.

Barton, H. (2000) Conflicting perceptions of neighbourhood. In: Barton, H. (ed.), *Sustainable Communities: The Potential for Eco-Neighbourhoods*. London: Earthscan.

Buehler, R. and Pucher, J. (2011) Sustainable transport in Freiburg: Lessons from Germany's environmental capital. *International Journal of Sustainable Transportation* 5(1): 43–70.

Bramley, G., Dempsey, N., Power, S., and Brown, C. (2006) What is 'social sustainability', and how do our existing urban forms perform in nurturing it. In: Sustainable Communities and Green Futures' Conference, Bartlett School of Planning, University College London, London.

Bramley, G., Dempsey, N., Power, S., Brown, C., and Watkins, D. (2009) Social sustainability and urban form: Evidence from five British cities. *Environment and Planning A* 41(9): 2125–2142.

Calthorpe, P. (1993) *The Next American Metropolis: Ecology, Community and the American Dream*. Princeton, NJ: Princeton Architectural Press.

Center for Transit-oriented Development [CTOD] (2009) *Mixed-Income Housing near Transit: Increasing Affordability with Location Efficiency*. Washington DC: Federal Transit Administration.

Consorcio Regional de Transportes de Madrid [CRTM] (2013) *Demanda de Transporte Público Colectivo* [*Public Transport Demand*]. Madrid: CRTM.

Daseking, W. (2015) Freiburg: Principles of sustainable urbanism. *Journal of Urban Regeneration & Renewal* 8(2): 145–151.

De Blas, M. (2004) ¿Quién quiere una ciudad? El plan parcial del sector PP5 del Arroyo Culebro, en Leganés [Who wants a city? The Arroyo Culebro PP5 sector master plan, in Leganés]. *Urban* 9.

Dempsey, N., Brown, C., and Bramley, G. (2012) The key to sustainable urban development in UK cities? The influence of density on social sustainability. *Progress in Planning* 77(3): 89–141.

Dempsey, N., Bramley, G., Power, S., and Brown, C. (2011) The social dimension of sustainable development: Defining urban social sustainability. *Sustainable Development* 19(5): 289–300.

Feldtkeller, A. (2015) Zwischen Planung und Selbstregelung. [Between planning and self-regulation] In: Scharf, A. and Gütschow, M. (eds), *Französisches Viertel* [*French Quarter*]. Tübingen: Stadtwandel, Regensburg.

Feldtkeller, A. (2017) Self-build neighbourhoods for living and working: A view from Tübingen, Germany. In: Benson, M. and Hamiduddin, I. (eds), *Self-Build Homes: Social Discourse, Experiences and Directions*. London: UCL Press.

Galster, G. (2007) Should policy makers strive for neighborhood social mix? An analysis of the Western European evidence base. *Housing Studies* 22(4): 523–545.

Galster, G.C. and Friedrichs, J. (2015) The dialectic of neighborhood social mix: Editors' Introduction to the Special Issue. *Housing Studies* 30(2): 175–191.

Hamiduddin, I. (2015a) The car in the neighbourhood: Residential design and social outcomes in southern Germany. In: Hickman, R., Bonia, M., and Banister, D. (eds), *Handbook of Transport*. Cheltenham: Edward Elgar.

Hamiduddin, I. (2015b) Social sustainability, residential design and demographic balance: Neighbourhood planning strategies in Freiburg, Germany. *Town Planning Review* 86(1): 29–52.

Hamiduddin, I. (2017) Journey to work travel outcomes from 'city of short distances' compact city planning in Tübingen, Germany [online]. *Planning Practice & Research* 23 October. https://doi.org/10.1080/02697459.2017.1378980.

Hamiduddin, I. and Gallent, N. (2016) Self-build communities: The rationale and experiences of group-build (Baugruppen) housing development in Germany. *Housing Studies* 31(4): 365–383.

Handy, S., Cao, X., and Mokhtarian, P. L. (2006) Self-selection in the relationship between the built environment and walking: Empirical evidence from northern California. *Journal of the American Planning Association* 72(1): 55–74.

Herzog, L.A. (2010) *Return to the Center: Culture, Public Space, and City Building in a Global Era*. Austin, TX: University of Texas Press.

Jenks, M. and Dempsey, N. (2007) Defining the neighbourhood: Challenges for empirical research. *The Town Planning Review* 78(2): 153–177.

Mokhtarian, P.L. and Cao, X. (2008) Examining the impacts of residential self-selection on travel behavior: A focus on methodologies. *Transportation Research Part B: Methodological* 42(3): 204–228.

Montgomery, R. (2016) State intervention in a post-war suburban public housing project in Christchurch, New Zealand. *Articulo – Journal of Urban Research* [online], 13. Available at: http://articulo.revues.org/2932.

Parker, T. and Arrington, G. E. (2002) *Statewide Transit-Oriented Development Study. Factors for Success in California. Final Report*. Business, Transportation and Housing Agency. California Department of Transportation.

Perry, C. A. (1929) City planning for neighborhood life. *Social Forces* 8(1): 98–100.

Sarkissian, W. (1976) The idea of social mix in town planning: An historical review. *Urban Studies* 13: 231–246.

Shirazi, M. R. and Keivani, R. (2017) Critical reflections on the theory and practice of social sustainability in the built environment–a meta-analysis. *Local Environment* 22(12): 1526–1545.

Universidad de Alcalá (2010) *Diagnóstico de Sostenibilidad, Agenda 21 [Sustainability Diagnosis, Agenda 21]*. Leganés: Leganés Town Hall.

Valenzuela, M. (1973) Los orígenes de los transportes urbanos y de cercanías en Madrid [The origins of urban transport and suburban trains in Madrid]. *Estudios Geográficos* 33(130): 95–132.

Wegener, M. (1994) Die Stadt der kurzen Wege – müssen wir unsere Städte umbauen [The city of short distances – We have to rebuild our cities], Faculty of Urban Planning Working Paper 136, University of Dortmund.

10 Social sustainability and collaborative housing

Lessons from an international comparative study

Richard Lang

Introduction

The aim of this chapter is to investigate the role of different international models of cooperative housing for social sustainability in the urban context. Although the focus is on the micro-scale of communities and neighbourhoods, the suggested analytical framework takes into account the multi-scalar and multi-level nature of social sustainability by introducing a vertical form of social capital. Although the phenomenon of cooperative housing has a long history, there is a recent wave of collective self-organised housing in many European countries, which emphasises aspects relevant to current challenges in urban development. These are discussed under the umbrella concept of sustainability, such as social inclusion and cohesion, environmental awareness, and affordability. Although there is no single unifying label that describes this emerging housing field, the chapter adopts 'collaborative housing' as an umbrella term that encompasses this wide variety of housing forms (Fromm, 2012; Vestbro, 2010). Its organisational models clearly exhibit cooperative principles (such as collective member ownership and democratic control) but nevertheless differ in important aspects from traditional and mainstream cooperative housing. They draw on new governance models and legal forms, and have sprung out of different social movements.

In contrast to for-profit or public providers, the ideal typical cooperative model draws on social capital as the key resource to build up and manage their organisations (Hatak *et al.*, 2016; Somerville and McElwee, 2011; Valentinov, 2004). Therefore, housing cooperatives have traditionally played a role in community development such as through engaging residents in local democratic and social economy practices (Birchall, 1988; Dawson, 1991; Mändle, 2013).

The empirical part of this chapter features new cooperative housing practice in two contrasting country contexts. First, given the longstanding history and strategic role of its cooperative housing sector, the case of Austria can provide lessons for the potential for collaborative models in other European contexts. In Vienna, social sustainability has become a central pillar of social housing policy that fosters social innovation in housing, including the promotion of new cooperative initiatives (Lang and Stoeger, 2018).

Second, a range of grassroots housing organisations is currently attempting to create a unified response to the highly dysfunctional and crisis-ridden housing market in England, heightened by the 2008 financial crisis. Under the label community-led housing, these organisations contribute proactively to the environmental agenda, for instance, through promoting carbon-light living arrangements. They represent alternative forms of housing provision based on resident self-organisation and community building, and got more political attention through recent moves towards localism (Heywood, 2016; Lang and Mullins, 2015; Mullins, 2012).

Although social sustainability appears to have become a powerful discourse in housing research and policy, a closer look at the concept reveals that it combines diverse goals, such as 'strong social networks' and 'social cohesion', but also the strife for 'social mixing' and 'social inclusion' in the neighbourhood (Williams, 2005: 146; Wohnfonds Wien, 2015). Thus, it represents a problematique, pointing to the opposite aspirations of belonging and differentiation in local communities (Novy, 2011). It is assumed that developing greater community cohesion, at the same time, generates social exclusion (Birchall, 1988; Williams, 2005). This ultimately leads to the question of the right balance between cohesion and diversity, which a city needs to be creative and innovative. In the light of this, the chapter explores how different collaborative housing models deal with contradictory principles of social sustainability.

The remainder of the chapter is structured as follows: the next section introduces the key theoretical concepts underpinning the analysis presented here – social sustainability, social capital and collaborative housing. After describing the methodological approach for this study, this chapter reports on the empirical findings of case studies for individual models of collaborative housing and explores the role of bonding, bridging, and linking social capital. As the social network is considered the core element of social capital (Brunie, 2009; Ferlander, 2007), this tripartite classification refers to types of relations among collaborative housing actors based on sociodemographic attributes as well as tie strength (Granovetter, 1973; Szreter and Woolcock, 2004). The final section provides a cross-case analysis and, in the concluding section, we reflect on the implications of the study findings for theory, policy, and practice of housing and social sustainability in the urban context.

Theoretical framework

Social sustainability and social capital in housing

The most common definition of sustainable development refers to the social, economic, and environmental pillars that were proposed by the Brundtland Commission which already mentions the contribution of housing (Hodge, 1997; World Commission on Environment and Development [WCED], 1987). Some scholars have added a fourth institutional or governance pillar (e.g. Pareja-Eastaway and Stoa, 2004). This classic sustainability concept has subsequently been applied and amended for urban housing research (e.g. Tosics, 2004; Winston, 2014; Winston and Pareja Eastaway, 2008) and for the field of collaborative housing (e.g. Coates, 2013; Delendi, 2017; Tummers, 2016). The economic dimension, for example, can refer to affordability of homes and lower financial costs in housing development, and the environmental dimension, for example, refers to efficient energy and water consumption as well as waste production within the building or housing scheme.

Although there is no consensus in the literature on defining the social dimension of sustainability, it is still possible to derive some key principles of the concept (Shirazi and Keivani, 2017). This chapter addresses two particular aspects of social sustainability: social inclusion, on the one hand, and social cohesion, on the other. Both aspects are highlighted by collaborative housing practice, but at the same time reflect an inherent contradiction in addressing different elements of social sustainability in housing.

To operationalise this problematique of different types of social relationships in urban communities and neighbourhoods, the chapter draws on social capital as a suitable analytical perspective located under the umbrella concept of social sustainability (Dale and Newman, 2010; Osborne *et al.*, 2016; Shirazi and Keivani, 2017). Despite the lack of an established definition, there is consensus among scholars that, at a generic level, the notion of social capital centres on social networks and embedded resources which can be mobilised through social interactions, leading to potential benefits for individual and collective actors (Brunie, 2009; Ferlander, 2007).

On the horizontal network level, an important distinction can be made between bonding and bridging social capital (Gittell and Vidal, 1998; Putnam, 2000). This distinction relates to boundaries of groups of residents on the neighbourhood level that share certain characteristics (Rydin and Holman, 2004). First, *bonding social capital* addresses the aspect of cohesion. It is considered an attribute of homogeneous networks and is also associated with trusting but inward-looking relationships that may constrain behaviour and flow of information (Granovetter, 1973; Poortinga, 2012). Thus, indicators for bonding capital in this study are connections of residents with similar sociodemographic characteristics, for example in terms of education, income, ethnicity, and social identity (including places), as well as frequent and intimate interactions among residents.

In contrast, *bridging capital* deals with the issue of social inclusion. It is associated with heterogeneous, less cohesive, and outward-looking relationships which are crucial to access new information and resources but also for one's integration into society (Granovetter, 1973; Poortinga, 2012). Indicators for bridging capital in this study are connections of residents with different sociodemographic characteristics, for example in terms of education, income, ethnicity, and social identity (including places), as well as less frequent and less intimate interactions among residents.

Finally, linking capital addresses the governance and institutional aspect of social sustainability, and highlights issues of power in social relationships. Similar to horizontal bridging capital, it 'cuts across different groups' (Ferlander, 2007: 119) but connects people vertically 'across power differentials' (Szreter and Woolcock, 2004: 655). Indicators for linking capital in this study are vertical ties between residents and institutional decision-makers, which can provide access to resources and information for the community of residents (e.g. legitimacy, consultancy, public funding) (Agger and Jensen, 2015; Jones, 2013; Lang and Novy, 2014). It has to be noted that linking capital, like every form of social capital, has negative effects when privileged access to resources leads to 'nepotism, corruption, and suppression' (Szreter and Woolcock, 2004: 655).

Previous research has highlighted the crucial role of intermediary institutional actors in building multi-level linking social capital (Agger and Jensen, 2015; Braunholtz-Speight, 2015; Lang and Novy, 2014). Intermediaries, such as collaborative housing umbrellas and secondary cooperatives, can support resident-led housing projects to access regime level actors, such as government bodies or large-scale housing providers, through coordinated outreach and mediation which helps leveraging local community resources.

In a nutshell, the three forms of social capital conceptualise the interplay of different types of social relationships and are thus relevant to urban sustainability. Social capital is a multi-dimensional concept and incorporates elements of both concepts, social cohesion and social inclusion, and further addresses the multi-level governance aspect in sustainable development (Dale and Newman, 2010; Osborne *et al.*, 2016; Ruiu, 2016).

Collaborative housing models and social capital

The traditional model of the community-based housing cooperative (Novy and Förster, 1991; Silver, 1991) can only partially be considered a good practice housing model that contributes to social sustainability in contemporary urban development. It is associated with high levels of bonding social capital, which typically lead to better mutual support among residents and a certain degree of community self-reliance. Such a cohesive community of residents is more capable of reacting to external changes and dealing with conflicts.

However, as a common-pool resource, bonding social capital requires institutional and governance support to be sustained over time. Unlike financial or human capital, it is a highly context-specific organisational asset (Ostrom, 1990). Furthermore, higher levels of bonding capital come at the expense of bridging capital in smaller community cooperatives as residents and management develop an inward orientation over time (Brandsen and Helderman, 2012; Lang and Novy, 2014). This trade-off between bonding and bridging capital is reinforced because collaborative housing initiatives often have a tendency towards self-selection of residents (Chiodelli and Baglione, 2014). In many cases, they are initiated by a core community of interest (e.g. around environmental awareness or an alternative lifestyle) and can lead to socially homogeneous resident structures. But, for social sustainability at the neighbourhood level, a certain level of bridging capital is important to address aspects of social inclusion and diversity (Forrest and Kearns, 2001).

Recent collaborative housing projects in Europe display new innovative approaches to developing multi-level linking social capital to powerful external resource holders in order to address the shortcomings of earlier waves of cooperative forms of housing. Such vertical linkages can be a way of facilitating both bonding and bridging social capital in urban neighbourhoods through access rules and regulation (Bovaird *et al.*, 2016; Jones, 2013).

Case study methodology

Collaborative housing is still a relatively under-researched phenomenon in both Austria and England. Thus, the empirical study reported here applied a qualitative research strategy with field work carried out between 2013 and 2017. The research draws on selected case studies from a large international research project on collaborative housing, which was carried out by the author. Qualitative multiple case studies appear suitable to reconstruct the meaning of collaborative housing in a concrete institutional and territorial context (Sayer, 1992; Yin, 2009). For this chapter, three collaborative housing models were selected in each country which show a clear link to the classical cooperative housing model and are relevant to the urban housing and development context in the respective country.

To contrast different actor perspectives, case studies draw on semi-structured interviews with managers/executives, resident representatives, and key external stakeholders, such as representatives of large housing providers or government bodies who contributed crucial resources to the case housing organisation. The interviews with managers and executives – who often also happened to be founding members of the initiative and residents – helped to reconstruct the history and governance structure of the case organisation. They also shed light on the linking social capital developed between the housing management and the community of residents, as well as on key external stakeholders. Semi-structured interviews with resident representatives were employed to gather complementary information on bonding and bridging social capital of residents, including connections with the external neighbourhood, as well as on interactions of residents with housing management. Interviews with key external stakeholders aimed to reveal how the case organisation built linking social capital with the housing regime level to gain legitimacy, funding, and other crucial resources. Wherever possible, the interviewees were encouraged to provide concrete examples of networking events, i.e. interactions of residents, housing management, and external stakeholders in key events of organisational life.

All interviews were taped and transcribed, and complemented by field observations of meetings and neighbourhoods. Therefore, the author spent several hours at case housing sites taking field notes of informal encounters with residents of the facilities and observations of interactions among staff members, management, residents, and neighbours. This extra effort helped to increase understanding of the case and thus the contextual and content-related plausibility of the data.

Primary data was complemented by archival data, such as that gathered from websites, reports, and media articles featuring the case organisations. This allowed the analysis to be based on the principle of triangulation, and to meet validity and reliability criteria of case study research. The content analysis (Mayring, 2010) helped to structure the empirical material gathered from interview transcripts, field notes, and documents. The category system was established deductively, reflecting the central analytical categories introduced in the previous section, i.e. social capital dimensions (bonding, bridging, and linking) and key actor groups involved in collaborative housing (e.g. residents, neighbours, housing management, regime level stakeholders). These categories were refined during the analysis. Details of the empirical sources referenced in this chapter are outlined in Table 10.1.

Discussion of results

Community-led housing models in England

In England, the term 'community-led housing' has been coined in practitioner discourses (Homes and Communities Agency [HCA], 2013) and has also been picked up by academics recently. The Housing Associations' Charitable Trust (HACT), for instance, refers to community-led housing as

> housing designed to meet the needs of particular groups of people or to meet the needs of a particular locality, and community-led, that is housing shaped and controlled by a group that represents the residents and/or the wider community that will be served by the housing.
>
> (Heywood, 2016: 12)

The community-led sector encompasses a range of organisational models with varying aims and resident target groups.[1] Table 10.2 provides a summary of the three English community-led housing models discussed in this chapter – cooperatives, co-housing, and community land trusts (CLT) – and their social capital characteristics. Although not explicitly linked to the cooperative housing tradition, co-housing and CLT also exhibit cooperative principles in their governance, such as community self-help, democratic member, or resident control and ownership (Lang and Mullins, 2015).

First, cooperatively owned housing is the longest-standing and largest sector within community-led housing in England with a focus on affordability, social inclusion, community cohesion, and resident participation. At present, there are about 836 housing cooperatives and mutual housing

Table 10.1 Overview of the empirical data

Acronym	*Case study*	*Housing model*	*Role of interviewee*	*Position of interviewee in case*	*Date of interview*
Interviews					
GCLT_ES	Granby Four Streets CLT	Community land trust (CLT)	Manager and board member of housing provider	External stakeholder (ES)	23 August 2016
GCLT_RR	Granby Four Streets CLT	Community land trust	CLT resident	Resident representative (RR)	24 August 2016
LCLT_ES	London CLT	Community land trust	Board member of community development charity	ES	4 October 2016
LCLT_MR	London CLT	Community land trust	Housing manager	Management representative (MR)	4 October 2016
NWHS_MR	North West Housing Services (NHWS)	Tenant-managed housing cooperative	Manager and board member of housing provider	MR	23 August 2016
RCH_ES	Redditch Co-operative Homes (RCH)	Tenant-managed housing cooperative	Local councillor	ES	14 December 2016
RCH_MR	RCH	Tenant-managed housing cooperative	Manager and board member of housing provider	MR	24 June 2013
SOVIESO_ES	so.vie.so	Participatory project	Housing consultant	ES	15 October 2013
SYNDICATE_MR	Willy*Fred	Syndicate model	Housing activist and project co-founder	MR	13 May 2014

continued

Table 10.1 Continued

Acronym	*Case study*	*Housing model*	*Role of interviewee*	*Position of interviewee in case*	*Date of interview*
Observation protocols (OP)					
ASPERN_OP	Seestern Aspern	Baugruppe	Guided tour of the housing scheme	Vienna, Austria	29 October 2015
LANC_OP	Lancaster Cohousing	Co-housing	Guided tour of the housing scheme	Lancaster, UK	26 June 2015
LILAC_OP	Low Impact Living Affordable Community (LILAC)	Co-housing	Guided tour of the housing scheme	Leeds, UK	17 April 2015
RCH_OP	RCH	Tenant-managed housing cooperatives	Resident viewing of new housing scheme	Redditch, UK	25 June 2016
SOVIESO_OP	so.vie.so	Participatory project	Guided tour of the housing scheme	Vienna, Austria	27 October 2015

Table 10.2 Social capital characteristics of case collaborative housing models in England

Collaborative housing model	*Bonding and bridging capital of residents*	*Linking capital between residents and external stakeholders*
Tenant-managed housing cooperatives	• Social housing tenants key target group, including households under stress • Residents have similar social backgrounds and share need for affordable homes • Schemes not specifically designed to encourage interaction with wider neighbourhood • Regular meetings that bring tenants from different cooperatives together	• Bottom-linked governance: mix of local resident self-management and top-down professional management by secondary cooperatives (housing associations) • Secondaries provide moderation, facilitation, and mediation for tenant communities in primary cooperatives • Secondaries facilitate access to finance, development, and management expertise for tenant groups • Tenants are directly represented on the board of secondaries • Longer-term partnerships and personal links between secondaries and local authorities help with access to land and funding
Co-housing	• Value-based intentional resident community with focus on environmental awareness • Collective home ownership and self-financing favours middle-class resident structure • Shared infrastructure and joint activities encourage daily resident interaction (e.g. common house and shared meals) • Some communal spaces open to externals from the local community (e.g. playground, co-working)	• Substantial bottom-up, resident self-management • Temporary support from external stakeholders, such as early stage grants from the Homes and Communities Agency or a social enterprise charity • UK Cohousing Network as loose umbrella organisation to support shared learning among co-housing projects, dissemination, and lobbying

continued

Table 10.2 Continued

Collaborative housing model	*Bonding and bridging capital of residents*	*Linking capital between residents and external stakeholders*
Community land trusts (CLT)	• Local residents looking for affordable homes who are committed to community participation • Wider local community involvement through open communal spaces and inclusive governance structures • Providing affordable housing only part of wider community service provision and urban activism	• Bottom-linked governance: mix of local community and resident self-management, and top-down professional management input from CLT umbrellas, larger housing associations, and civil society intermediaries (e.g. for campaigning, community, and capacity building) • Good alignment with government discourse and priorities around housing and communities • Successful in attracting early stage public grants • Partnerships with larger housing providers or local council entail challenge of marrying contrasting institutional logics

organisations across the UK, managing a stock of around 169,000 units, with 91% of cooperatives based in England (Heywood, 2016). Our two case studies of housing cooperatives are Redditch Co-operative Homes (RCH) in the West Midlands and North West Housing Services (NWHS) in Liverpool.

As so-called 'secondary' cooperatives, RCH and NWHS service a range of tenant-managed cooperatives ('primaries') in the locality, which in the case of RCH make up altogether about 300 homes, whereas NWHS services 30 'primaries' ranging from 12 residents to 220 residents (NWHS_MR). Most of these housing cooperatives have been formally incorporated within the social housing sector, meaning that they are registered providers, regulated by the HCA and eligible for government funding. As a management representative of NWHS points out, the social backgrounds of residents are fairly similar: 'they're working class communities; so people who would otherwise probably be in other types of social, rented housing'. Residents also come from the local council's waiting list (RCH_MR), including households under stress and in urgent housing need, as a housing manager highlights in relation to a particular project: 'they came from the clearance site and moved on to the new site where the co-op has developed. The majority of our clients' background is from that angle' (NWHS_MR).

Each cooperative is run by a tenant management committee that is elected from and by the membership. The secondary management is primarily in personal contact with committee members, although it services all tenants and acts as an agent who does not want to interfere with resident self-governance. Its focus is on supporting the volunteer committee with professional expertise in tenancy management, including legislation and financing, and dealing with external stakeholders, such as the HCA or the local council (NWHS_MR). Tenants from 'primaries' are directly represented on the board of the secondary cooperatives (RCH_MR). RCH and NWHS managers facilitate community building among the tenants early in the planning process, keep in regular personal contact with tenants, and establish clear mechanisms for conflict resolution (RCH_MR).

Cooperative schemes are not necessarily designed to encourage interaction with the wider neighbourhood, but tenants from different coop schemes get to know each other at regular meetings and some tenants even swap homes (RCH_OP). Furthermore, RCH established a longstanding partnership with the local council, which facilitates access to building land and funding streams (RCH_ES). Similarly, NWHS developed vertical personal linkages to key local decision-makers, including local authority professionals and local council leadership. In return for public funding, some of the residents are nominated from council waiting lists (RCH_ES).

Second, co-housing comes close to the traditional cooperative model with its aims to establish an intentional community of residents, limited in size, and based on mutual support and shared values. In 2015, there were 19 completed co-housing schemes in England (Heywood, 2016). Our case projects for this model are Lancaster Cohousing and LILAC (Low Impact Living

Affordable Community) in Leeds, which emerged in the 2000s. Residents lead the design, planning, maintenance, and management of their schemes, which should be tailored to encourage a strong sense of community. One main difference from cooperative housing is the importance of shared infrastructure in co-housing schemes. In our case organisations, we find a common house with kitchen and dining room where residents cook together and share meals, a laundry, a food cooperative, a workshop, a car club, and some shared outdoor spaces, which should altogether make the individual resident's everyday life more efficient and encourage daily face-to-face interaction with neighbours.

Nevertheless, both case projects are keen to open their communal spaces to externals and the wider local community, such as through a playground or a co-working and event space (LANC_OP, LILAC_OP). The two cases are also characterised by architectural design elements, which are innovative in the territorial context, such as a Passivhaus construction (LANC_OP) and the use of straw and timber as natural building materials (LILAC_OP). Besides reducing the environmental impact, this underlines their character as a distinct housing community within the wider neighbourhood. The cases show that co-housing can appeal to different age groups and lifestyles, with older single residents and younger families being neighbours.

The case projects foreground financial self-management, intentional and values-based living, environmental awareness, and collaborative work–living arrangements. Although co-housing schemes can exist in different tenure types, the cases are typical in that they focus on collective forms of home ownership that aim to counter some of the dysfunctions of mainstream home ownership (LANC_OP, LILAC_OP). They have been successful in drawing on early stage grant funding, such as from a social enterprise charity, the government's Homes and Communities Agency, and some initial support from the local council (Chatterton, 2013). Both case projects are also members of the UK Cohousing Network, which represents a rather loose umbrella organisation for co-housing projects. It supports mutual learning among the schemes and dissemination activities, including government lobbying, together with other community-led housing umbrella organisations.

Third, about 160 CLT projects have become effective vehicles through which negotiations take place to dispose public land assets to community groups around England (Thompson, 2015). CLTs have recently gained relevance through their ability to address rural poverty and local market overheating associated with over-concentrations of second homes. In contrast to the cooperative movement, CLTs are primarily set up for local community benefits rather than just member or resident benefit. Furthermore, they have managed to develop multi-level linking strategies that align well with current government discourse and priorities around housing and communities. Individual CLT projects are supported by a network of regional umbrella organisations that link up to a national umbrella. Although most projects have been set up in rural areas (e.g. the St Minver CLT in Cornwall), there has been a new wave of urban CLTs, such as our case projects, the East London CLT and the Granby Four Streets CLT in Liverpool, addressing the urgent need for more affordable housing in cities.

The specific challenge for the East London CLT involved high land prices and plots too small for real community development. This led to the strategy of including affordable housing parcels in planning permissions. Building on the community-organising experience and campaigning efforts by Citizens UK around the issue of the high cost of local housing, the London CLT gained political legitimacy through establishing direct personal linkages to local and central government actors, including interactions with successive mayors of London (LCLT_ES). The importance of linking social capital is highlighted by the following statement of a management representative of the London CLT: 'it starts with a pledge from the local Mayor or local Councillors to commit some time or the land [...]. Citizens UK run actions where they'll go and present a cardboard house or something to a Councillor and it's all about building those relationships on an institutional level and getting noticed and then developing them' (LCLT_MR). Vertical networking helped to access a suitable site and early stage grant money, which together with private subsidies supports feasibility and project development, although for a rather small number of 23 affordable homes for sale for local families at the CLT's pilot project at St Clement in Mile End. Drawing on input from Citizens UK and local community leaders, the London CLT management organises the allocation process, develops resident self-management skills, and facilitates community building and the use of communal spaces, which should bring together residents and local community. CLT membership is open to anyone living in London and the board should equally represent residents, local community, and stakeholders (LCLT_MR).

The case of Granby Four Streets CLT is different from the London CLT project in that an existing scheme was part of a former Housing Market Renewal area that saw years of decline and low land values. A group of residents resisted the demolition of their houses and instead wanted to refurbish them and keep their local community (GCLT_RR). A CLT member recalls how a neighbouring couple initiated the residents' association: '"[l]ook, the Council want to knock your houses down. What are we going to do about it?" And that's how that started 25 years ago.' Although, from a professional view, it would have been easier to finance the rebuilding, the CLT model eventually turned out to be most supportive in meeting the aspirations of the group. As in most other CLTs, the case involved partnerships with external stakeholders, including NHWS, which bring in important professional expertise and facilitation. In the Granby case, this also involved mediating the interests of the local council and the local resident group. A management representative of NHWS describes the impact of their external support on the Granby CLT as follows: 'Eventually, they've got the capability now to take empty properties off the Council and refurbish them as well as creating employment opportunities and facilities in the area, like shops and training'

(GCLT_ES). Furthermore, the architects who worked with the group during the refurbishment process eventually won the Turner Prize, which gave this CLT project a very high profile. Nevertheless, both the Granby and the London CLT projects also show that partnerships with external stakeholders, such as local councils or larger housing providers, are challenging in terms of marrying contrasting institutional logics (GCLT_ES; see also Moore, 2016).

Collaborative housing models in Austria

Collaborative housing (*gemeinschaftliches Bauen und Wohnen*) in Austria postulates an alternative to the pronounced top-down approach of mainstream social housing which limits resident influence in planning and housing management (Lang and Novy, 2014). Our analysis focuses on the local context of Vienna where most of current collaborative housing activity takes place. In the given institutional environment, affordable collaborative models can hardly be initiated without external support from either the municipality or large limited-profit housing[2] associations. Community groups generally lack the financial capacity to compete against corporate investors on the free market. At present, there are two main models of realising collaborative housing within the realm of subsidised housing and one on the private market. These models and their socially sustainable potential are discussed in the following paragraphs. Table 10.3 provides a summary of the three Austrian collaborative housing models discussed in this chapter – participatory projects, *Baugruppen*, syndicate – and their social capital characteristics.

Many well-established large housing associations in Vienna still use the label 'cooperative' although they have moved away from genuine resident participation and democratic governance principles (Lang and Novy, 2014). In contrast to those mainstream cooperatives, our first case project 'so.vie.so'[3] (an acronym for 'Sonnwendviertel Solidarity') represents an emerging top-down model of collaborative housing in the social housing sector where tenants start participating in a predesigned structure about three years before the actual completion of the scheme, facilitated by architects and external consultants. The scheme was realised in 2013 and is part of a large integrated neighbourhood development (5,000 homes for about 13,000 residents), focusing on social cohesion and local access to infrastructures (such as public transports, schools, health care. etc.) (SOVIESO_OP).

The participatory project 'so.vie.so' consists of 111 subsidised rented apartments, with a tenant's advisory board and tenant-managed communal facilities of different sizes (e.g. rooftop garden, workshop, or gym), and shared greenspace with the adjacent housing schemes. In contrast to mainstream social housing schemes in Vienna, residents thus enjoy increased opportunities for participation and are given responsibility, which is highlighted by an external consultant for this housing project:

Table 10.3 Social capital characteristics of case collaborative housing models in Austria

Collaborative housing model	*Bonding and bridging capital of residents*	*Linking capital between residents and external stakeholders*
Participatory model	• Mixed and middle-class resident structure • Resident nomination by provider and local council according to (initial) commitment to community participation • Tenant-managed communal facilities of different sizes (e.g. rooftop garden, workshop, or gym) • Shared greenspace with adjacent housing schemes	• Top-down model of collaborative housing within the subsidised social housing framework • Externally facilitated participation process and community building • Case scheme is part of integrated neighbourhood development informed by social sustainability criteria (e.g. mixed providers and tenures) • Tenant's advisory board for governance of the scheme • Residents have limited influence on design of scheme and provider governance
Baugruppen	• Intentional core community of residents based on shared lifestyle, including interest in neighbourhood development • About a third of residents nominated from council • Some communal space open to externals (e.g. co-working)	• Bottom-linked governance: mix of local resident self-management and top-down professional management by a housing association • Realised within subsidised housing framework of the local authority • Case schemes part of an integrated neighbourhood development plan • Residents have substantial influence on planning and development process
'Syndicate model'	• Group of younger activists and residents interested in collective affordable living • Residents share values of strong solidarity and self-help • Semi-public communal spaces (e.g. workshop, courtyard) • Host for other local civil society initiatives as well as workshops, discussion groups, etc.	• National umbrella association similar to a secondary cooperative is key shareholder in local housing projects • Collective private ownership model that is financed through bank loan and crowd-funding model • No links to local authority or housing associations

> usually you have residents who would like to engage but do not find opportunities [...] although house rules and tenancy law clearly regulate the rights of an individual tenant, they constrain any bottom-up initiative from the residents [...] thus, the goal of this project was to enable resident participation in practice.
>
> (SOVIESO_ES)

Financing for this collaborative model mainly comes from direct housing subsidies, which are secured by a large limited-profit housing association in a developer competition for the specific site. In return the provider guarantees affordable rents and high design quality. Both provider and municipality select and nominate tenants for the scheme from their housing waiting lists. Together with the fairly generous income ceilings in social housing in Vienna, this leads to a certain social mix of residents, including singles, families, and also residents with migrant backgrounds. However, for this project, the selection focus was in addition on residents willing to contribute to community development. Furthermore, residents pay an initial one-off contribution to the building costs which is usually between 15.000 and 30.000 euro for a medium-sized to large flat (SOVIESO_OP).

The *Baugruppe* is the second Austrian collaborative model discussed here. It can broadly be defined as a housing project that is (co-)initiated, (co-) planned, and (co-)constructed by future residents. In addition, it often aims to create an intentional community (Temel *et al.*, 2009). Typical case projects have recently been developed in the Viennese neighbourhood Aspern Seestadt, which is based on an integrated urban development concept, such as the projects 'Pegasus' and 'Seestern Aspern' (see also Hendrich, 2012; Lang and Stoeger, 2018). 'Pegasus', for instance, consists of 25 housing units of different sizes with a shared rooftop and basement communal areas, and a children's playing area. This model is also realised within the subsidised housing framework of the local authority in Vienna. The key difference to the participatory projects similar to 'so.vie.so' is that *Baugruppen* are initially constituted by a resident group, organised in an independent association, which teams up with a limited-profit housing association as the developer. In a *Baugruppe* like 'Seestern Aspern', which has 27 flats, the resident group enjoys more possibilities of participation in the planning of the scheme, which in turn can lead to higher construction costs for the architect and developer. Typically, the resident group individually rent two-thirds of the flats from the housing association and collectively decide on their re-allocation. For the remaining flats, the municipality reserves the right to nominate residents. This can be regarded as a potential conflict for community building and the governance of this collaborative model. The analysed *Baugruppen* target residents of different ages who are creative people interested in taking an active role in community and neighbourhood development. Thus, the projects are also keen to open some communal spaces to externals, such as through a co-working space in the project 'Seestern Aspern' (ASPERN_OP).

Finally, the 'Syndicate model' has emerged on the collaborative housing scene in different Austrian cities in the last a few years. The target group of the model is rather younger residents interested in collective affordable living in urban locations who share values of solidarity and self-help (SYNDICATE_MR). Although it is already well developed in Germany under the label 'Mietshäuser Syndikat' ('tenant syndicate'), it recently managed to realise its first project called 'Willy*Fred' in the Austrian city of Linz, with an existing house consisting of 14 flats of different sizes and semi-public communal spaces, such as a workshop, courtyard, and rehearsal room for bands. The housing project also hosts other local civil society initiatives as well as workshops, discussion groups, etc. More Syndicate projects are currently under way in Austria, one in Salzburg, three in Vienna, and one in Innsbruck.[4] These resident-led local housing projects are developed in the legal form of a limited company within the Austrian umbrella association 'Dachverband habiTAT'. The umbrella can be regarded as a secondary cooperative that supports the individual local projects as a 51% shareholder, and thus also guarantees its non-profit status. Each individual project forms a tenant association that operates according to grass-roots democratic principles and holds the remaining 49% of the limited company. In our case of 'Willy*Fred', the association represents 20 tenants who form small volunteer groups to manage and maintain the building. The 'Willy*Fred' group in Linz bought its house from a private owner with a bank loan that was largely secured through a direct crowd-funding loan. The group is constantly seeking new 'investors' to keep the model going (F Humer, 2018, personal communication).

Cross-case analysis

Evidence across all cases presented in this chapter suggests that the creation of bonding social capital – understood as community cohesion based on more intimate relationships and mutual help among residents – represents the core of collaborative housing projects. There is, however, another aspect of bonding capital, related to similarities in residents' social characteristics, which is important when analysing the relationship between these housing models and social sustainability.

According to the case results, residents of cooperatives in England often share the same background as social housing tenants looking for community-led alternatives in the affordable housing segment. In contrast, co-housing groups in England, as well as *Baugruppen* and the Syndicate model in Austria, mostly represent intentional communities where residents share a certain alternative lifestyle. Nevertheless, the resident communities of these three housing models also want to stimulate bridging social capital by proactively reaching out to the external neighbourhood and to other community initiatives, which confirms findings from previous empirical studies (e.g. Ruiu, 2016). However, the case evidence from England and Austria presented in this chapter suggests that the ability of private collective ownership models to foster social inclusion on a larger scale is limited compared with models that are embedded in social and affordable housing programmes. This comes down to the crucial role played by multi-level linking social capital.

In English and Austrian cases, we find larger non-profit associations (including secondary cooperatives) that facilitate communities' access to affordable land and funding through their links with the local authority. In the cases of English cooperatives and urban CLTs, social housing tenants are not only provided with access to affordable collaborative living but also accommodated with management skills. Thus, linking capital can help disadvantaged resident groups to 'get on' in life more generally instead of just 'getting by'.

Compared with cases in England, the approach taken by the local authority in Vienna provides a more comprehensive framework of sustainable urban development in which smaller-scale *Baugruppen* and the larger participatory projects are embedded. Crucial to this are the social housing developer competitions with a set of criteria explicitly related to social sustainability. This makes sure that any subsidised housing project has to consider resident participation and community building. New urban development also consists of housing with different tenures and subsidisation to ensure a social mix that goes beyond intergenerational living to mixed income groups in neighbourhoods. In Vienna, the underlying logic is social mixing in neighbourhoods that the local authority proclaims as a key element in securing social stability throughout the city. Furthermore, there is the approach of integrated neighbourhood development, which altogether provides a suitable framework for urban collaborative housing projects (Lang and Stoeger, 2018). Another example of embedding collaborative housing and *Baugruppen* in integrated neighbourhood development is the Vauban district in Freiburg, in Germany (Coates, 2013).

Nevertheless, the key challenge of embedding collaborative models in social housing programmes – whether in Austria or in England – is that groups of future residents have to develop their concept of a community within a given framework. The inherent hypothesis is that communities can be created by specific means of architectural and social design, but residents in collaborative housing also need a degree of autonomy to experiment with community and social capital building. The English CLT cases in this chapter show that umbrella bodies and secondary cooperatives can provide crucial mediation between conflicting institutional logics of community groups and professional housing management, as well as government professionals. In the case of *Baugruppen*, but also in CLT projects, the nomination of social housing tenants through the local authority and larger providers can, however, cause potential conflicts with an established intentional resident community.

Conclusions and implications for research, policy, and practice

The chapter investigates the role of different international models of collaborative housing for social sustainability in the urban context. It aims to offer a better understanding of the concept and practice of social sustainability in an emerging European housing sector that is based on collective self-organisation of residents. Thus, it also addresses the claim for more international comparative and place-based empirical investigation into the still under-researched concept of social sustainability (Boström, 2012; Kytta *et al.*, 2016; Shirazi and Keivani, 2017). The qualitative case analyses focus on exploring the contradictory relationship between two key principles of social sustainability in the urban sphere: cohesion and inclusion. On the scale of communities and neighbourhoods, both aspects are highlighted by collaborative housing practices. To examine their complex interrelationship empirically, the chapter mobilises the tripartite approach to social capital.

Linking social capital – representing the vertical form of social capital – can be seen as an informal governance for bonding and bridging relationships in urban neighbourhoods. The case analyses suggest that collaborative housing models display a range of vertical linking strategies that connect resident communities with powerful external stakeholders. In this respect, the chapter highlights the crucial role of multi-level linkages and intermediary actors who mediate and translate between resident groups and housing regime actors, such as large non-profit associations and local authorities. However, more comparative research is needed on the configuration of linking social capital in different places and housing models, including types of intermediaries, and its interplay with bonding and bridging capital.

The analysis further suggests that collaborative models developed within subsidised housing programmes and integrated neighbourhood development can address both the cohesion and the inclusion aspects of social sustainability. This leads to greater financial accessibility of collaborative housing schemes for different sociodemographic groups, and it also offers physical and non-physical infrastructures to create bridges to the external neighbourhood. Housing developer competitions for collaborative projects, on plots reserved for them, has emerged as a successful path in Vienna in recent years. However, collaborative housing that is developed in this way can prosper only as long as there is a political will to support it. Moreover, any institutional support – be it from the local authority or a large housing provider – needs to respect the organisational autonomy of resident groups. Otherwise there is the danger that collaborative housing projects and their social capital will ultimately be co-opted by the top-down logics of urban development programmes (Gressgård, 2015).

Collaborative housing practices, as well as policy strategies discussed in this chapter, are inherently place based. Thus, international transferability of the results outlined in this chapter depends on similarities of institutional conditions, such as the availability of comparable legal forms. However, the Syndicate and *Baugruppen* models show that ideas and elements can indeed be transferred between countries, as is the case between Austria and Germany. The same is true for community land trusts in England and Belgium (Aernouts and Ryckewaert, 2017) and the spread of the co-housing model to various European countries. Besides its distinct cooperative nature, the Austrian 'Syndicate model' also shows similarities to the English CLT movement with broader ideological aims of turning property into community ownership with an asset lock on resale, thus counteracting increasing private speculation (Moore, 2016).

In principle, the developer competition model of Vienna can also be applied anywhere as a promotional tool for social sustainability. It probably has the potential for application to countries characterised by a certain degree of federalism or devolution in housing. However, longitudinal case studies are needed to deliver more insights into the effects of different organisational and institutional models of collaborative housing for advancing social sustainability in the urban context.

Acknowledgements

The author is grateful to Ernst Gruber and David Mullins for their helpful comments, which helped to improve the article.

This work was supported by an APART-fellowship of the Austrian Academy of Sciences [Project number 11696] and by a Marie Curie Intra-European Fellowship (IEF) [Project number 622728].

Notes

1 For a comprehensive overview of community-led housing models in England see Lang and Mullins (2015) and Heywood (2016).

2 Limited-profit housing (*gemeinnütziger Wohnbau*) is a dominant type of tenure in social housing in Austria. It is provided mainly by large private associations and cooperatives which are limited in the distribution of their profit shares. This housing tenure is further based on the principles of affordability and cost rent. Provision is reliant on public subsidies and favourable urban land policies, and targets a relatively wide range of household income levels (Lawson, 2010; Ludl, 2007).

3 Thanks to Cora Strogach for interview data collection that also informed this case study.

4 See https://habitat.servus.at/ (accessed 12 April 2018).

References

Aernouts N and Ryckewaert M (2017) Beyond housing: On the role of *commoning* in the establishment of a Community Land Trust project. *International Journal of Housing Policy*. DOI: 10.1080/19491247.2017.1331592.

Agger A and Jensen JO (2015) Area-based initiatives – And their work in bonding, bridging and linking social capital. *European Planning Studies* 23(10): 2045–2061.

Birchall J (1988) *Building Communities the Co-Operative Way*. London: Routledge & Kegan Paul.

Boström M (2012) A missing pillar? Challenges in theorizing and practicing social sustainability: Introduction to the special issue. *Sustainability: Science, Practice, & Policy* 8(1): 3–14.

Bovaird T, Stoker G, Loeffler E, Jones T, and Roncancio MP (2016) Activating collective coproduction for public services: Influencing citizens to participate in complex governance mechanisms. *International Review of Administrative Science* 82(1): 47–68.

Brandsen T and Helderman JK (2012) The trade-off between capital and community: The conditions for successful co-production in housing. *Voluntas* 23(4): 1139–1155.

Braunholtz-Speight T (2015) Scottish community land initiatives: Going beyond the locality to enable local empowerment. *People, Place and Policy* 9(2): 123–138.

Brunie A (2009) Meaningful distinctions within a concept: Relational, collective, and generalized social capital. *Social Science Research* 38(2): 251–265.

Chatterton P (2013) Towards an agenda for post-carbon cities: Lessons from Lilac, the UK's first ecological, affordable cohousing community. *International Journal of Urban and Regional Research* 37(5): 1654–1674.

Chiodelli F and Baglione V (2014) Living together privately: For a cautious reading of cohousing. *Urban Research & Practice* 7(1): 20–34.

Coates GJ (2013) The sustainable urban district of Vauban in Freiburg, Germany. *International Journal of Design & Nature and Ecodynamics* 8(4): 265–286.

Dale A and Newman L (2010) Social capital: A necessary and sufficient condition for sustainable community development? *Community Development Journal* 45(1): 5–21.

Dawson J (1991) Housing cooperatives: From housing improvement to economic regeneration? *Local Economy: The Journal of the Local Economy Policy Unit* 6(1): 48–60.

Delendi ML (2017) Sustainable living models. *International Journal of Sustainable Development and Planning* 12(4): 772–779.

Ferlander S (2007) The importance of different forms of social capital for health. *Acta Sociologica* 50(2): 115–128.

Forrest R and Kearns A (2001) Social cohesion, social capital and the neighbourhood. *Urban Studies* 38(12): 2125–2144.

Fromm D (2012) Seeding community: Collaborative housing as a strategy for social and neighbourhood repair. *Built Environment* 38(3): 364–394.

Gittell R and Vidal A (1998) *Community Organizing: Building Social Capital as a Development Strategy*. Thousand Oaks, CA: Sage.

Granovetter M (1973) The strength of weak ties. *American Journal of Sociology* 78, 1360–1380.

Gressgård R (2015) The power of (re)attachment in urban strategy: Interrogating the framing of social sustainability in Malmö. *Environment and Planning A* 47(1): 108–120.

Hatak I, Lang R, and Roessl D (2016) Trust, social capital, and the coordination of relationships between the members of cooperatives: A comparison between member-focused cooperatives and third-party-focused cooperatives. *Voluntas* 27(3): 1218–1241.

Hendrich P (2012) [ro*sa]. In: id22: Institute for Creative Sustainability: Experimenticity (ed.), *CoHousing Cultures: Handbook for Self-Organized, Community-oriented and Sustainable Housing*. Berlin: Jovis, pp. 132–147.

Heywood A (2016) Local housing, community living: Prospects for scaling up and scaling out community-led housing. London: The Smith Institute. Available at:

www.smith-institute.org.uk/book/local-housing-community-living-prospects-for-scaling-up-and-scaling-out-community-led-housing (accessed 1 February 2017).

Hodge T (1997) Towards a conceptual framework for assessing progress toward sustainability. *Social Indicators Research* 40(1–2): 5–98.

Homes and Communities Agency (HCA) (2013) Community-led development: Eligibility criteria. Available at: www.gov.uk/government/uploads/system/uploads/attachment_data/file/394037/community_led_eligibility.pdf (accessed 1 February 2017).

Jones T (2013) Community capital and the role of the state: an empowering approach to personalisation. *People, Place and Policy* 7(3): 153–167.

Kyttä M, Broberg A, Haybatollahi M, and Schmidt-Thomé K (2016) Urban happiness: Context-sensitive study of the social sustainability of urban settings. *Environment and Planning B: Planning and Design* 43(1): 34–57.

Lang R and Mullins D (2015) Bringing real localism into practice through co-operative housing governance. The role and prospects for community-led housing in England. HCRG Working Paper 1-2015, Housing and Communities Research Group (HCRG), University of Birmingham, UK. Available at: www.birmingham.ac.uk/Documents/college-social-sciences/social-policy/IASS/housing/2015/working-paper-series/HCR-WP-1-2015.pdf (accessed 1 February 2017).

Lang R and Novy A (2014) Cooperative housing and social cohesion: The role of linking social capital. *European Planning Studies* 22(8): 1744–1764.

Lang R and Stoeger H (2018) The role of the local institutional context in understanding collaborative housing models: Empirical evidence from Austria. *International Journal of Housing Policy* 18(1): 35–54.

Lawson J (2010) Path dependency and emergent relations: Explaining the different role of limited profit housing in the dynamic urban regimes of Vienna and Zurich. *Housing, Theory and Society* 27(3): 204–220.

Ludl H (2007) Limited-profit Housing Associations in Austria. Vienna: Austrian Federation of Limited-Profit Housing Associations (GBV). Available at: www.iut.nu/members/Europe/West/Austria_GBVinfo_ENG.pdf (accessed 5 April 2018).

Mayring P (2010) *Qualitative Inhaltsanalyse. Grundlagen und Techniken* [*Qualitative Content Analysis. Basics and Techniques*], 11th edn. Weinheim: Beltz Verlag.

Mändle M (2013) The economics of housing co-operatives externalities. *International Journal of Co-Operative Management* 6(2): 25–32.

Moore T (2016) Replication through partnership: the evolution of partnerships between community land trusts and housing associations in England. *International Journal of Housing* 18(1): 82–102.

Mullins D (2012) The Localism Act and reform of social housing. In: Raine J and Staite C (eds), *The World Will Be Your Oyster? Perspectives from the Institute of Local Government Studies (INLOGOV) on The Localism Bill.* Birmingham: Institute of Local Government Studies, University of Birmingham, pp. 89–99.

Novy A (2011) Unequal diversity – On the political economy of social cohesion in Vienna. *European Urban and Regional Studies* 18(3): 239–253.

Novy K and Förster W (1991) *Einfach bauen: Genossenschaftliche Selbsthilfe nach der Jahrhundertwende: Zur Rekonstruktion der Wiener Siedlerbewegung* [*Building Simple: Co-Operative Self-Help after the Turn of the Century. Reconstructing the Viennese Settlers' Movement*]. Vienna: Picus.

Osborne C, Baldwin C, and Thomsen D (2016) Contributions of social capital to best practice urban planning outcomes. *Urban Policy and Research* 34(3): 212–224.

Ostrom E (1990) *Governing the Commons: The Evolution of Institutions for Collective Action*. Cambridge: Cambridge University Press.
Pareja-Eastaway M and Stoa E (2004) Dimensions of housing and urban sustainability. *Journal of Housing and the Built Environment* 19(1): 1–5.
Poortinga W (2012) Community resilience and health: The role of bonding, bridging, and linking aspects of social capital. *Health & Place* 18(2): 286–295.
Putnam RD (2000) *Bowling Alone: The Collapse and Revival of American Community*. New York: Simon & Schuster.
Ruiu ML (2016) The social capital of cohousing communities. *Sociology* 50(2): 400–415.
Rydin Y and Holman N (2004) Re-evaluating the contribution of social capital in achieving sustainable development. *Local Environment* 9(2): 117–133.
Sayer A (1992) *Method in Social Science: A Realist Approach*. London: Routledge.
Shirazi MR and Keivani R (2017) Critical reflections on the theory and practice of social sustainability in the built environment: A meta-analysis. *Local Environment* 22(12): 1526–1545.
Silver H (1991) State, market, and community: Housing co-operatives in theoretical perspective. *Netherlands Journal of Housing and the Built Environment* 6(3): 185–203.
Somerville P and McElwee G (2011) Situating community enterprise: A theoretical exploration. *Entrepreneurship and Regional Development* 23(5–6): 317–330.
Szreter S and Woolcock M (2004) Health by association? Social capital, social theory, and the political economy of public health. *International Journal of Epidemiology* 33(4): 650–667.
Temel R, Lorbek M, Ptaszyńska A, and Wittinger D (2009) *Baugemeinschaften in Wien: Endbericht 1, Potenzialabschätzung und Rahmenbedingungen [Building Cooperatives in Vienna: Assessment of Potential and Conditions]*. Vienna: City of Vienna, MA50.
Thompson M (2015) Between boundaries: From commoning and guerrilla gardening to community land trust development in Liverpool. *Antipode* 47(4): 1021–1042.
Tosics I (2004) European urban development: Sustainability and the role of housing. *Journal of Housing and the Built Environment* 19(1): 67–90.
Tummers L (2016) The re-emergence of self-managed co-housing in Europe: A critical review of co-housing research. *Urban Studies* 53(10): 2023–2040.
Valentinov V (2004) Toward a social capital theory of cooperative organization. *Journal of Cooperative Studies* 37(3): 5–20.
Vestbro D (ed.) (2010) *Living Together – Co-Housing Ideas and Realities around the World*. Stockholm: Royal Institute of Technology Division of Urban Studies in collaboration with Kollektivhus NU.
Williams J (2005) Sun, surf and sustainable housing – Cohousing, the Californian experience. *International Planning Studies* 10(2): 145–177.
Winston N (2014) Sustainable communities? A comparative perspective on urban housing in the European Union. *European Planning Studies* 22(7): 1384–1406.
Winston N and Pareja Eastaway M (2008) Sustainable housing in the urban context: International sustainable development indicator sets and housing. *Social Indicators Research* 87(2): 211–221.
Wohnfonds Wien (2015) Beurteilungsblatt 4-Säulen Modell [Assessment form 4-pillar model]. Available at: www.wohnfonds.wien.at/media/file/Neubau/4-Saulen-Modell.pdf (accessed 27 May 2016).
World Commission on Environment and Development (WCED) (1987) *Our Common Future*. Oxford: Oxford University Press.
Yin RK (2009) *Case Study Research: Design and Methods*. Thousand Oaks, CA: Sage.

11 Social sustainability as an urban agenda

Towards a comprehensive image

M. Reza Shirazi and Ramin Keivani

The present and future of urban social sustainability research

As outlined and discussed in Chapter 1, social sustainability discourse faces some challenges in areas of theory and evaluation. We begin the concluding chapter underlining the contribution of this book to these challenges.

Contributors to this volume confirm that social sustainability, compared with other aspects of sustainable development, has not been sufficiently discussed and explored in the field they are studying. Looking at the academic literature produced during the recent years, a large number of which has been referenced in this book, shows that the literature on social sustainability is progressing, though at a slower pace than other dimensions (Shirazi and Keivani, 2017, 2018). We would argue that urban social sustainability discourse has moved from an 'under-theorised' status to an 'insufficiently theorised' status; we have a considerable body of knowledge produced on this topic, but more theoretical and empirical research needs to be conducted to consolidate the discourse and advance its theory. This book, taking into account the multidisciplinary nature of the chapters, provides an overview on multiple, but not all, aspects of urban social sustainability discourse, and moves it some steps further.

The 'ambiguity of definition', we would suggest, is no longer interpreted as a threat and disadvantage, but as an opportunity and asset for the social sustainability discourse. First, as we have discussed in detail elsewhere (Shirazi and Keivani, 2017), although social sustainability has been approached fragmentally and presented itself as a 'concept in chaos', a closer look suggests that some key features and characteristics construct the very foundation of the concept such as: equity; democracy, participation, and civic society; social inclusion and mix; social networking and interaction; livelihood and sense of place; safety and security; human well-being; and quality of life. Moreover, as Davison puts it in Chapter 2, social sustainability appears as an 'empty signifier': it is a term with flexibility that admits multiple but relevant content. It is this flexibility that helped social sustainability to serve as an umbrella concept and take advantage of similar but long-standing and well-developed social concepts such as social equity, social justice, social cohesion, and quality of life, as we outlined in Chapter 1. This makes social sustainability structurally and dialogically interconnected to other old, new, or emerging social conceptualisations. In this way, social sustainability provides a constructive dialogical arena, and invites other valuable social ideas to participate, negotiate, exchange, and debate. It presents itself as a bridge language, a *lingua franca*, to establish, maintain, and promote dialogue and communication. Furthermore, flexibility of conceptualisation paves the way for flexibility of operationalisation: social sustainability is defined, conceptualised, and operationalised to address specificities of the given context. In this sense, we would argue, disregarding social sustainability for being vague and slippery appears to neglect the potentiality of being vague and slippery, and can lead to institutional lock-in, and stifle discourse and limit innovations in policy and practice.

However, we should not underestimate the practical implications of such an approach: If multiple meaning can be attached to social sustainability, what content should be included and which formulation should serve as a departure point for policy-making? This is an important political question which brings us to the question of decision-making, consensus building, and governance, as discussed later.

This volume covers a wide range of disciplines in the built environment, and thus provides a 'multidisciplinary image' from urban social sustainability. It tries to demonstrate that social sustainability is relevant to any urban intervention and (re)development scheme, and that planning or designing the built environment contains elements of social sustainability. This book covering multiple fields such as urban governance, mobility, architecture, urban space, urban heritage, urban policy, eco-urbanism, neighbourhood planning, and housing shows that urban social sustainability is multidisciplinary and multifaceted. This collection brings different disciplines together, but certainly does not cover all possible ones: there are still several disciplinary areas in the built environment in which the social sustainability dimension remains unexplored, or is in early stages of development, such as urban health, architecture (interior space), zero-carbon urbanism, etc.

Despite covering different disciplines, this volume only has one case from the Global South that is under-represented in social sustainability research. Broadening geographical coverage of social sustainability discourse, to gain a more context-related and nuanced understanding from the social dynamics of cities and societies, remains a gap that needs to be filled in future studies. Diversifying social sustainability research in terms of geographical coverage, supported by empirical data from the field, could also push the limits of the 'normativity' of the social sustainability concept and bring more analytical input into the discourse. As noted in Chapters 1 and 2, unlike environmental sustainability, which refers to tangible and verifiable environmental facts, social sustainability moves quickly into normative territory so that qualities and criteria suggested for improving social sustainability standards are contested, open to interpretation, and place specific, with no consistency. What is 'the' better social condition towards which we should move is a key question of urban social sustainability; the answer to this hardly moves away from being essentially normative.

As many of the chapters in this volume are case study research, they suggest suitable evaluation frameworks, employing qualitative and quantitative methods, to measure social sustainability in different fields, and thus contribute to the operational challenges of social sustainability discourse. The proposed frameworks (e.g. Chapters 4, 5, 6, 7, 9, and 10) are designed to select the most relevant indicators, and use appropriate data collection techniques. This volume respects multi-scalarity of social sustainability and covers different scales from the regional down to the local, and the content follows a clear spatial order. However, in line with a general shift from the macro- to the micro-scale in social sustainability research (Shirazi and Keivani, 2017), 'the local' in different shapes appears as a scalar focus for the authors. In Chapter 3, Tunström highlights the importance of the neighbourhoods in Nordic cities for planning of society, as places where principles of inclusive urban space should be respected and implemented. Hamiduddin and Adelfio (Chapter 9) discuss the question of scale in urban planning discipline and argue that the neglect of neighbourhood as a physical and social entity reflects the broader underdevelopment of social sustainability discourse in both research and policy. Netto, Vargas, and de Saboya (Chapter 7) find an assemblage of urban blocks as the best scale to investigate material conditions of social sustainability in urban spaces. Holden, Airas, and Larsen in Chapter 8 focus on urban neighbourhoods to analyse societal challenges and contradictions of eco-urbanism, as do Shirazi and Keivani in Chapter 6 to map outdoor social activities. Lang (Chapter 10) looks at a smaller scale of collaborative housing projects. The scalar selection of study area for researching urban social sustainability has two implications: first, scale of inquiry should be carefully selected to serve as the best context for the analysis; and, second, social sustainability is scale-less in the sense that every urban scale has a social sustainability dimension eligible for investigation.

Holden, Airas, and Larsen (Chapter 8) explain how the geography of operation matters for social sustainability. They argue that, for decision-making and implementation of policies and projects, the conflict between social sustainability values as a global concept and the Nordic welfare system as a deep-seated regional model appears as a dilemma:

> This dilemma comes into focus in cities in the Nordic region, where efforts to create social sustainability predate the use of the term 'social sustainability' in international policy dialogue, but the new international urban policy landscape imposes a different register of values and risks, costs and benefits of sustainability efforts against which cities are expected to perform.
>
> (p. 152)

They show how model-sustainable projects and practices, such as eco-neighbourhoods, exacerbate inequity, social segregation, and uneven development when exported to other places with different sociocultural and political environments. They conclude that 'If eco-urban neighbourhood developments aim to create prototypes suitable for building "anywhere" they lose the ability to carry connections to local contexts including local political settings, values, cultures and identities' (p. 165). Avoiding generalisations and mass export–import industry of ideas and models, and taking into account characteristics of the given urban setting, is a priority for the practice of social sustainability models.

Although this volume contributes to the theory and practice of urban social sustainability discourse, as briefly noted above, future research should be mobilised to address more intensively the core challenges it faces: conducting more comparative, empirical studies which (1) investigate cases from different sociocultural and geographical settings, with mixed qualitative and quantitative methods, and multiple territorial scales from macro to micro; (2) establish a constructive dialogue with similar social concepts (see Chapter 1); (3) expand disciplinary coverage to include new areas of inquiry and develop under-researched fields; and (4) develop place-specific, subject-related frameworks for evaluation and measurement.

Policy challenges of urban social sustainability

Social sustainability is not exclusively a subject for theorisation as a scientific topic, nor is it solely a framework for empirical research purposes, but an area of concern for urban policy and development. Policy-makers and governmental bodies have developed urban policies, planning strategies, and social programmes to enhance social sustainability of cities and communities. However, this has never been an easy task.

Urban social sustainability is fundamentally related to urban policy. As Davidson (2009) puts it, seeking a socially sustainable urban future requires an engagement with policy-making, and thus not only opens up the space of politics, but also places social politics at the centre of the debate. Any type of planning for urban futures is inevitably a question about social sustainability, because, as noted in Chapter 1, social sustainability remains an integral component of any sustainable urban future. This is why social sustainability is receiving greater attention in some municipalities worldwide, as we can find in policy documents such as 'Policy Report: Definition of Social Sustainability' (Vancouver), 'Social Sustainability Strategic Plan' (Boulder, Colorado), 'Burnaby Social Sustainability Strategy' (Burnaby, British Colombia), 'Social Sustainability Vision' (Fort Collins, Colorado), and 'Social Sustainability Policy and Action Plan 2018–2028' (Sydney). In the remainder of this chapter, we discuss key challenges of social sustainability with regard to urban policy, and drawing on them formulate the main argument of the book, which suggests that urban social sustainability should be treated as an urban agenda.

The first policy challenge of urban social sustainability is 'thematic negligence': social sustainability is not recognised as an integral part of urban planning and urban policy, and has rarely been integrated into future plans, urban initiatives, and urban governance structure. For example, a survey conducted in the US shows that the social aspect of sustainability is the least addressed dimension in local governments' sustainability activities, initiatives, and plans. Although 94% of local governments state that 'economy' has a very high priority and 62% give priority to the 'environment', only 38% consider 'social justice' as an important factor (Svara *et al.*, 2014). Portney (2013), looking at US cities, argues that most of the cities have not incorporated social sustainability principles such as social justice and environmental equity into their sustainability initiatives and plans, and those that have done so have done it in only superficial ways: 'Perhaps it would be more accurate to say that as practiced in most cities [of the US], equity issues do not appear to be integral parts of cities' definition of sustainability' (Portney, 2013: 204).

Some contributors to this volume note that any urban project has a social sustainability dimension that needs to be taken into consideration from the beginning. For example, Uteng, Singh, and Hagen (Chapter 4) point out that the first step in designing a transport project to be socially sustainable is integrating principles of social sustainability, such as affordability, availability, and safety, from the early stages of design. Similarly Landorf (Chapter 5), in her analyses of four World Heritage urban landscape nomination documents and management plans, argues that none of them satisfactorily integrates social sustainability principles into the heritage planning and management process.

The second challenge is 'lack of transformative methodology': the problem of developing appropriate methodologies to incorporate social sustainability concerns into urban development planning. Turner (2012), in his study of the gender equity in a mass transport system in Jakarta, argues that mainstreaming of gender into planning is a fundamental component of delivering social sustainability. However, the transport planning profession has failed to produce systematic methodologies that can incorporate gender issues into urban planning and development. This is not a challenge of urban transportation, but is applicable to many other fields, and appears as a critical challenge for policy-makers. In their study of two partnerships in health-promotion initiatives to improve food security in the State of Victoria, Australia, which covered 12 municipalities, Slade and Carter (2017) argue that, although these initiatives aim at integrating some elements of social sustainability, a systematic method was missing. 'Local governments were constrained in their capacity to respond to all aspects of social sustainability, and their prioritisation of equity, or indeed any social goal, may be a strategic choice when undertaking new initiatives in constrained circumstances' (Slade and Carter, 2017: 396). For the World Heritage plans, as Landorf notes in Chapter 5, developing adequate performance indicators is a main challenge. She argues that this is more critical when intangible aspects such as equitable fairness, inclusion, diversity, and quality of life are to be measured and evaluated.

The third challenge is 'legitimising instrumentality'. Urban policies addressing social sustainability have been criticised for exploiting social sustainability rhetoric for justifying, beatifying, and legitimising re-development programmes, and propagating new ways of marketing mechanisms with wide negative social impacts on the disadvantaged and marginalised citizens. For example, Lees (2014) reviews the influential presence of social mix policies in British housing policy and argues that social sustainability in London 'is merely rhetoric used by local authorities to sell the wholesale demolition of council estates and their "regeneration" as mixed communities. The end result is the displacement of low-income populations from inner London, state-led gentrification and social polarisation/segregation' (Lees, 2014: 150). Instrumentalising social sustainability for legitimising urban policies and projects, or for marketing purposes, threatens the politics of urban social sustainability and endangers credibility of such policies and projects for the citizens.

Finally, the way in which social sustainably criteria are formulated, defined, and incorporated into urban policy should not be the outcome of a 'non-dialogic procedure', but result from a collaborative and inclusive social environment. In some cases, incorporating social sustainability into strategic planning and development documents has been criticised for being less about serving as a practical means for equalising living conditions and more about mobilising people under a future vision through establishing a fantasmatic narrative. Gressgård (2015) critically reflects on the 'Overview Plan for Malmö 2012' strategic plan and challenges its inclusivity as being projected from a single point of view, despite asserting dialogue and citizen participation. In fact, integrating social sustainability into urban policy is a complicated and sophisticated project, and faces many other serious challenges such as division of local government roles and responsibilities, coordination of different sectors, scales of operation, level and nature of partnership between agents, and institutional capacity of local authorities in terms of human and financial resources. It is also important, as mentioned, to develop methodologies that secure engagement of a wide spectrum of citizens and take into consideration their priorities and problems.

Advocating a dialogical procedure for integrating social sustainability into urban policy is rooted in the principal of 'democracy', which, as we have suggested (Shirazi and Keivani, 2017), is a constituent element of urban social sustainability. Linking democracy to urban social sustainability has multiple implications, a main one of which is that urban social sustainability should be interpreted as a political project, elements of which are negotiated and developed through a legitimised political process. Democracy, for sure, is a contested concept, but seeks for equal access for all members of society to mechanisms of decision-making, achieved through an inclusive political process. As Tunström notes (Chapter 3), integrating communicative and participatory tools into planning processes at the local and city scale, providing opportunities for public involvement, and active civil society forums provide a platform for developing socially sustainable initiatives. Landorf (Chapter 5) finds out that, in the World Heritage management plans, the particular concern for social sustainability is the persistence of limited community participation. Despite the inclusion of government and non-government agencies in the planning process, actual and effective community involvement is very limited and scarce. She argues that 'The active involvement of a broad selection of stakeholders also endangers a collective sense of ownership and commitment not found through consultation on already developed concepts and strategies' (p. 94).

Before we go further and suggest how these challenges could be addressed, in the next section we point out two critical attributes of urban social sustainability that inform urban policy.

Spatiotemporal relativism and procedural action

Urban social sustainability is a relational concept in terms of place and time. First, what is socially sustainable in one particular place could prove to be unsustainable in another context. This indicates that urban social sustainability discourse should be moderated by context and locality. This moderation takes place in terms of definition, measurement, and policy-making. Although there are key themes and concepts on which urban social sustainably discourse is based, such as equity and democracy, these concepts should be moderated for the specific context, and relevant indicators for evaluation purposes should be tailored. This geographical relativism problematises the question of knowledge exchange and urban policy transfer: model social sustainability policies and projects should be treated carefully in wider circulation and transfer to other sociopolitical contexts.

Second, what is socially sustainable now may prove to be unsustainable in the future. This character reflects the very dynamic nature of societies as they consistently evolve and transform. The temporal relativity of urban social sustainability becomes problematic and challenging when it comes to planning and programming. Development of a flexible structure that accommodates probable changes over time and adapts itself to sociopolitical and cultural advances, alterations, and distortions is a difficult task. Temporality of urban social sustainability does not, however, deny 'planning' and 'programming', but underlines the significance of 'flexible planning': making planning frameworks flexible enough to accommodate temporal changes of societies.

The recognition of temporal dynamics of urban social sustainability has a significant consequence: we are always on the journey. The 'socially sustainable condition' does not exist, but is a 'changing path' towards 'socially more sustainable condition'. This is why urban social sustainability necessitates a procedural action: 'In reaching for social sustainability, we are engaged in a social process, rather than working toward a clear and unchanging goal' (Dillard *et al.*, 2009: 290). There are some definitions that look at social sustainability as a product and end state. For example, Yoo and Lee (2016) argue that social sustainability is an outcome of social capital, and thus is about an end state: 'Social sustainability can be defined as an ability to sustain society, because "sustain" means maintaining a certain state that is determined by both physical and nonphysical factors' (Yoo and Lee, 2016: 4), and then 'social sustainability is defined as the ability to maintain one society's certain state or existence' (Yoo and Lee, 2016: 6). However, the dominant perception underscores the procedural nature of urban social sustainability. For example, McKenzie's definition that social sustainability is 'a life-enhancing condition within communities, and a process within communities that can achieve that condition' (McKenzie, 2004: 12) emphasises the processes through which a community moves towards a better condition. Boström (2012) distinguishes between substantive and procedural dimensions of social sustainability: Although a substantive dimension addresses the 'what' of social sustainability, namely the principal qualities of a socially sustainable future, the procedural dimension addresses the 'how' of this shift and move, namely the means for achieving those substantive qualities.

Urban social sustainability, we would argue, is not exclusively about preserving and sustaining existing societal qualities for the present and future generations, but enhancing, improving, and upgrading them. To be sure, societies inherit social qualities that they would like to sustain them and carry them to the next generation. However, societies have never been, and are not, in an ideal social condition. This dialectic of 'preserving' and 'changing', or 'sustaining' and 'improving', is not a binary opposition, requiring an 'either–or' approach, but shows two sides of the urban social sustainability coin. In other words, an important part of the urban social sustainability agenda is mitigating and alleviating social unsustainability. In this sense, we see urban social sustainability as a process with two seemingly opposite but actually supplementary components of 'preservation' and 'transformation'. It is about neither an absolute status quo nor a radical transfiguration, but a process of relatively slow and continuous shift towards a better future. Urban social sustainability does not imply stability; a socially sustainable society is not a socially stable society, but one in which social sustainability criteria transform to reflect the dynamic nature of society.

We refer to this dialectical relationship as 'politics of pathway': this recognition that any urban social sustainability agenda should draw on an in-depth 'understanding' of 'status quo' and a realistic 'projection' of an 'ideal condition'. It clarifies the nature of the avenue, which should culminate in a better future. The politics of pathway is the art of acting within these dialectics, and guides us to the key questions of 'governance'.

Politics of pathway and governance of urban social sustainability

As we discussed in Chapter 1, the social dimension is integral to urban sustainability; what matters is the way constituent components and defined pillars of sustainable development interact and relate to each other. Generally speaking, two types of relationship could be assigned to the linkage between sustainable development pillars: self-contained and interconnected. In the self-contained approach, pillars of sustainable development are sufficiently independent and equally important that they can be treated separately, so that each pillar could be defined, conceptualised, and operationalised in isolation from other constituent elements. This approach may have practical value for policy-making purposes, but faces critical theoretical challenges such as being reductionist and over-simplified. The interconnected approach suggests that, regardless of the number of pillars we define for sustainable development, all of them are equally important, but structurally interrelated. This approach legitimises developing concepts, frameworks, policies, and strategies for each pillar, provided that we bear in mind, and address appropriately, their interconnectivity with other pillars in an arena of dialogue, negotiation, and interaction. Here, we advocate the interconnected approach. We suggest that urban social sustainability should be treated as a valuable, semi-independent concept, interconnected to other elements of sustainability, so that developing an urban social sustainability agenda remains a valid policy action, taking into account the relative ignorance of the subject in sustainable development theory and practice.

If urban social sustainability is an open-ended process, as we suggested above, the governance of urban social sustainability gains significance, because an open-ended process also requires an efficient governance system that covers the entire process from planning and design to implementation and monitoring. Governance is a pivotal domain for societal transformation towards sustainable development (Lange *et al.*, 2013). Sustainability governance focuses on processes through which sustainability principles are incorporated into political decision-making mechanism (Pettibone, 2015). It is important to know which type of governance is more suitable to achieve urban sustainability goals. Governance for sustainable development 'implies the deliberate adjustment of practices of governance in order to ensure that social development proceeds along a sustainable trajectory' (Meadowcroft *et al.*, 2005: 5). It deals with the ways by which the society can be steered towards an overarching objective that embraces social, economic, and political goals now and into the future (Baker, 2009).

The paradigm shift from government to governance indicates that, in order to achieve social sustainability goals and objectives, the governance of urban social sustainability should ensure effective coordination and coherent action of all the actors, including civil society and formal governmental bodies. Thus, good governance necessitates involving and incorporating all stakeholders into the decision-making process in order to increase political engagement (Evans *et al.*, 2006). To promote socially sustainable development, collaborative, deliberative, and polycentric governance modes are advocated which encourage partnership among multiple actors and establish relationships among various policy levels (Lange *et al.*, 2013). Following

Kemp *et al.* (2005, p. 26), we can argue that, in order to progress towards higher social sustainability,

> we need to establish governance structures and practices that are able to foster, guide and coordinate positive work by a host of actors on a vast complex of issues, through webs of interconnection and across multiple levels and scales, with sensitivity to their contexts and respect for uncertainties.

An active and dynamic governance system that promotes cooperation with civil society is capable of achieving social sustainability goals. Studies show that a greater civic engagement with, and trust in, the local government policy-making and action leads to higher sustainability achievements (Evans *et al.*, 2006). What Kemp *et al.* (2005) suggest as key features and components for the governance of sustainability is also relevant to the governance of urban social sustainability: policy integration that could be achieved through vertical policy integration within the governmental sector and horizontal policy integration across policy sectors; shared objectives; imposing sustainability-based criteria for planning and approval of significant undertakings; developing explicit rules and processes for decisions about trade-offs and compromises; identification of place-specific and integrated sustainability indicators; developing relevant means and tools to spur and guide appropriate action; and promoting system innovation to secure long-term improvements.

An important aspect of urban social sustainability governance addresses the dialectics of 'preserving' and 'changing'. The idea that any urban social sustainability agenda should take into consideration what qualities should be preserved and what should be improved necessitates a process of constructive dialogue and communication between different social groups, namely citizens and community members. In other words, this should be achieved through a democratic process of communication, negotiation, and decision-making. In fact, an integral part of governing urban social sustainability is establishing a democratic platform through which members of the community, a neighbourhood, city, or region, identify essential qualities of a socially sustainable society, and define what they want to preserve and what they want to change. However, as Grieco (2015) states, identifying relevant stakeholders and agents to be involved in decision-making processes for urban social sustainability remains a critical question.

It is also important to identify the most appropriate scale for governing urban social sustainability. The question is, what is the more practical scale for governing and managing urban social sustainability? According to Stren and Polese (2000), social theory has dealt with issues of social change at a macro-level, ignoring the day-to-day realities of urban life that take place at the local level. They argue that much of the social sustainability-related issues are seemingly prosaic matters at the very local scale, such as design of streets and the removal of garbage, and this underpins the significance of local policies and local institutions, but bearing in mind the dynamism and complexity of urban change. However, the administrative and managerial limits of the local institutions should not be underestimated, because they may lack resources and motivation to act independently (Frisken *et al.*, 2000). We would suggest that urban social sustainability is multi-scalar, and so is planning for it; all the scales from a small housing project up to a regional development plan are potential arenas for addressing social sustainability. In other words, planning for urban social sustainability should occur at multiple scales and territories, provided that a horizontal correspondence and relationship between agencies are secured.

It is beyond the scope of this concluding chapter to theorise 'governance of urban social sustainability'. However, we would like to close this section by reviewing four challenges of urban social sustainability related to urban policy, and explore how the governance of urban social sustainability should address these challenges.

To address the 'thematic negligence' of urban social sustainability in urban policy we need to recognise that urban social sustainability is an integral part of any urban planning initiative. It raises this awareness that any urban development initiative, from local to regional and beyond, has a social sustainability dimension that must be properly addressed. In other words, recognising 'urban social sustainability' as an essential component of 'urban sustainable development' is the first step, in order to move from 'thematic negligence' to a 'thematic recognition' condition. We have evidence to show that in some cases this thematic recognition is already in place, and has led to a number of initiatives. A good example is the case of 'social sustainability commissions', as Tunström notes in Chapter 3. In Sweden, these commissions, initiated by cities such as Malmö, Gothenburg, and Stockholm, seek to develop strategies for mitigating social challenges of segregation and social injustices, and health inequalities.

As Rashidfarokhi *et al.* (2018: 1288) put it, a problem that prevents the integration of social sustainability themes into planning processes is the 'lack of knowledge and understanding of the ways to integrate social sustainability into the processes'. The lack of 'transformative methodology' in urban social sustainability is the question of operation after recognition: developing frameworks to understand status quo, and suggesting methodologies through which the existing condition can be transformed into a better future. In other words, transformative methodology is about developing a pathway, and this is a main task of governance of urban social sustainability. A 'dialogical procedure', rooted in the principle of 'democracy', is the most appropriate way to define the 'pathway'. As we noted, the governance of urban social sustainably should be inclusive and collaborative. As Manzi *et al.* (2010: 221) put it, operation of urban social sustainability 'requires nested, multi-layered institutional arrangements for decision-making that are consistent in intent, coherent in approach and collaborative in diversity. These arrangements need to be carefully planned, equitably provided and continually monitored at community, neighbourhood and city-wide levels'. Thus, governing urban social sustainability requires establishing a dialogical decision-making procedure through which all relevant agents collaborate to define the social sustainability pathway.

Urban social sustainability as an urban agenda

Approaching urban social sustainability as an urban agenda, we would argue, is inspired by 'socially aware planning' (Baldwin and King, 2018) and is rooted in this recognition that everything urban has a social sustainability dimension that should be carefully and comprehensively addressed. Everything urban, in the context we are discussing, points to a range of areas such as urban policies, strategic plans, future plans, (re)development projects, local initiatives, housing developments, etc. This does not mean an over-emphasis on social sustainability dimension, or isolating it from other dimensions, but calls for a dialogue with other aspects of sustainable development.

Urban social sustainability as an urban agenda has three fundamental elements: recognition, integration, and monitoring. It starts with recognition: recognising that there is a social sustainability dimension for every (re)development project or programming that has to be respected and planned for. Regardless of the scale of operation, this recognition paves the way for planning and programming. It does not take place in a vacuum, but in a social arena where multiple agents are engaged, and pushed by societal demands and public awareness. The other two elements of integration and monitoring are part of the governance of urban social sustainability. Governance of urban social sustainability is circular (Figure 11.1); it starts with a dialogical planning

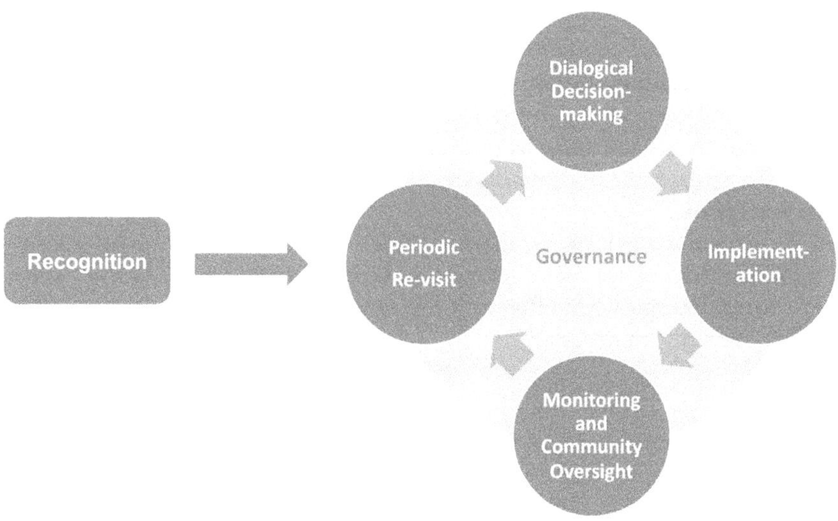

Figure 11.1 Approaching urban social sustainability as an urban agenda.

and design, goes further with implementation, and continues with participatory monitoring (with a focus on community oversight). The circle closes with periodic re-visit, which urges reforms or if needed re-structuring in planning, design, or even monitoring mechanism. The re-revisit includes evaluation of decision-making, planning, design, and implementation procedures to evaluate the achievements. In larger scales, for example urban policies that are long term and unlimited, the circle repeats itself several times. It necessitates a horizontal integration to link actors, stakeholders, and institutions, as well as a vertical integration to interconnect discrete levels of governance from local to regional and beyond.

The main features of approaching urban social sustainability as an urban agenda are as follows:

- Addressing urban social sustainability acknowledges interconnectivity of social sustainability with other pillars, establishes a constructive dialogue with them, and makes suggestions for integration, exchange, and interaction with them.
- It seeks institutional integration so that urban administration places urban social sustainability at the core of its governance structure.
- There is a territorial multi-scalarity in the sense that urban social sustainability operates at multiple scales, from the local to the regional and beyond.
- It is a process-oriented action/activity; it does not remain in 'status quo' but moves towards a future vision where social sustainability qualities improve.
- It applies to multiple fields from housing to transport and health.
- It emerges out of a collective, collaborative, decision-making process, engaging all the possible stake-holders in the phase of preparation, implementation, and monitoring.
- It acknowledges community oversight in all stages from design to implementation and evaluation.
- It is self-critical and self-reflexive, because it periodically re-visits its achievements and suggests reforms for betterment. This reflects dynamics of societies and is receptive to societal changes of values, behaviours, and perceptions.
- It favours integration; in more complex cases or at larger urban scales, a horizontal linkage with relevant institutions, agents, and actors, and a vertical linkage with different levels of governance from local to regional and beyond are necessary.

Ultimately, it is the task of policy-makers to develop needed institutional and professional capacities and introduce policy regulations that require social sustainability appraisal and action framework for different urban (re)development plans, schemes, and programmes. This will be realised only if this urgency is first recognised. The framework we proposed above can serve as a generic framework for all (re)development urban projects and initiatives.

References

Baker S (2009) In pursuit of sustainable development: A governance perspective. In: 8th International Conference of the European Society for Ecological Economics, Ljubljana.

Baldwin C and King R (2018) *Social Sustainability, Climate Resilience and Community-Based Urban Development: What about the People?* New York: Routledge.

Boström M (2012) A missing pillar? Challenges in theorizing and practicing social sustainability: introduction to the special issue. *Sustainability: Science, Practice, & Policy* 8(1): 3–14.

Davidson M (2009) Social sustainability: A potential for politics? *Local Environment* 14(7): 607–619.

Dillard J, Dujon V, and King MC (eds) (2009) *Understanding the Social Dimension of Sustainability*. New York: Routledge.

Evans B, Joas M, Sundback S, *et al.* (2006) Governing local sustainability. *Journal of Environmental Planning and Management* 49(6): 849–867.

Frisken F, Bourne LS, Cad G, *et al.* (2000) Governance and social sustainability: The Toronto experience. In: Polese M and Stren R (eds), *The Social Sustainability of Cities: Diversity and the Management of Change*. Toronto: University of Toronto Press, pp. 68–97.

Gressgård R (2015) The power of (re)attachment in urban strategy: Interrogating the framing of social sustainability in Malmö. *Environment and Planning A* 47: 108–120.

Grieco M (2015) Social sustainability and urban mobility: Shifting to a socially responsible pro-poor perspective. *Social Responsibility Journal* 11(1): 82–97.

Kemp R, Parto S, and Gibson RB (2005) Governance for sustainable development: Moving from theory to practice. *International Journal of Sustainable Development* 8(1–2): 12–30.

Lange P de, Driessen PPJ, Sauer A, *et al.* (2013) Governing towards sustainability: Conceptualizing modes of governance. *Journal of Environmental Policy and Planning* 15(3). Available at: http://dspace.library.uu.nl/handle/1874/280842 (accessed 9 August 2018).

Lees L (2014) The death of sustainable communities in London? In: *Sustainable London? The Future of a Global City*. Bristol: Policy Press, pp. 149–171.

Manzi T, Lucas K, Jones TL, *et al.* (eds) (2010) *Social Sustainability in Urban Areas: Communities, Connectivity and the Urban Fabric*. London, Washington, DC: Routledge.

McKenzie S (2004) *Social Sustainability: Towards Some Definitions*. Hawke Research Institute Working Paper Series 27. Magill: Hawke Research Institute.

Meadowcroft J, Farrell KN, and Spangenberg J (2005) Developing a framework for sustainability governance in the European Union. *International Journal of Sustainable Development* 8(1–2): 3–11.

Pettibone L (2015) *Governing Urban Sustainability: Comparing Cities in the USA and Germany*. Farnham: Ashgate.

Portney KE (2013) *Taking Sustainable Cities Seriously: Economic Development, the Environment, and Quality of Life in American cities*, 2nd edn. Cambridge, MA: The MIT Press.

Rashidfarokhi A, Yrjänä L, Wallenius M, *et al.* (2018) Social sustainability tool for assessing land use planning processes. *European Planning Studies* 26(6): 1269–1296.

Shirazi MR and Keivani R (2017) Critical reflections on the theory and practice of social sustainability in the built environment: A meta-analysis. *Local Environment* 22(12): 1526–1545.

Shirazi MR and Keivani R (2018) The triad of social sustainability: Defining and measuring social sustainability of urban neighbourhoods. *Urban Research & Practice* 1–24.

Slade C and Carter J (2017) Local governance for social sustainability: Equity as a strategic response to neoliberal constraints in food security initiatives. *Australian Geographer* 48(3): 385–399.

Stren R and Polese M (2000) Understanding the new sociocultural dynamics of cities: Comparative urban policy in a global context. In: Polese M and Stren R (eds), *The Social Sustainability of Cities: Diversity and the Management of Change*. Toronto: University of Toronto Press, pp. 13–38.

Svara JH, Watt T, and Takai K (2014) *Local Governments, Social Equity, and Sustainable Communities. Advancing Social Equity Goals to Achieve Sustainability*. Washington, DC: ICMA.

Turner J (2012) Urban mass transit, gender planning protocols and social sustainability: The case of Jakarta. *Research in Transportation Economics* 34(1): 48–53.

Yoo C and Lee S (2016) Neighborhood built environments affecting social capital and social sustainability in Seoul, Korea. *Sustainability* 8(12): 1346.

Index

Page numbers in **bold** denote tables, those in *italics* denote figures.

For Product Safety Concerns and Information please contact our EU representative GPSR@taylorandfrancis.com
Taylor & Francis Verlag GmbH, Kaufingerstraße 24, 80331 München, Germany

www.ingramcontent.com/pod-product-compliance
Ingram Content Group UK Ltd.
Pitfield, Milton Keynes, MK11 3LW, UK
UKHW011003180926
475701UK00004B/532
9780367662035